PREDICTING THE
ECLIPSE
A MULTIMILLENNIUM TALE OF COMPUTATION

PREDICTING THE
ECLIPSE
A MULTIMILLENNIUM TALE OF COMPUTATION

STEPHEN WOLFRAM

Predicting the Eclipse: A Multimillennium Tale of Computation

Copyright © 2024 Stephen Wolfram, LLC

Wolfram Media, Inc. | wolfram-media.com

ISBN-978-1-57955-087-5 (paperback)
ISBN-978-1-57955-088-2 (ebook)

Science/Astronomy

Cataloging-in-publication data available at wolfr.am/Eclipse-cip

Typeset with Wolfram Notebooks: wolfram.com/notebooks

CONTENTS

Wolfram website for precision eclipse computation:
www.precisioneclipse.com

Preface

There's a certain drama—and surprise—to a total solar eclipse. Suddenly the Sun that so predictably crosses the sky every day is gone—covered by a perfect black disk. It's hard to even imagine what the ancients must have thought. For them, the heavens, and things like the motion of the Sun and Moon, were a beacon of precision and perfection in an otherwise seemingly random and hard-to-predict world.

Probably in ancient times no single person ever saw more than one total eclipse in their lifetime. But no doubt they wondered whether there would be another eclipse, and when it would be. And indeed, "When will the next eclipse be?" may well be the longest-running precise question ever. And to answer it took a chain of science, mathematics and computation that's basically spanned the whole history of human civilization. And tracing the story of this process provides a unique window into intellectual history.

I had long known the broad outlines of the history of predicting eclipses. But as I came to research it in more detail in preparation for the 2017 eclipse, I was amazed at just how many prominent scientists and mathematicians had been involved over the centuries. More than I knew, eclipse prediction was a microcosm—and driver—of exact science. By the early twentieth century much had been achieved. But the final hero of the story—as in so many modern stories of science—was computation. It was—to paraphrase Isaac Newton—"beyond the force of any human mind" to work out the detailed motions of the bodies in our solar system. But with enough effort, computers could do it, and eventually they did.

And in 2017 we mounted a demonstration: a website that successfully predicted to within one second the time when the eclipse would arrive at any particular location. Even then it was somewhat challenging to achieve this. But in the few years since, we've advanced to the point where now the necessary computations have become a completely routine matter in the Wolfram Language.

In this book I have tried both to tell the story of the intellectual adventure that led to the successful prediction of eclipses, and to describe the science, mathematics and computation that was involved in it. I don't assume specialized technical knowledge, though even experts will probably find things they were not familiar with. But more than anything else, my goal is to give a sense of a great intellectual achievement that involved so much and spanned so many centuries—and finally gave us the triumph of modern eclipse prediction.

Stephen Wolfram

When Exactly Will the Eclipse Happen? A Multimillennium Tale of Computation

Preparing for April 8, 2024

On April 8, 2024, there's going to be a total eclipse of the Sun visible on a line across the US. But when exactly will the eclipse occur at a given location? Being able to predict astronomical events has historically been one of the great triumphs of exact science. But how well can it actually be done now?

The answer is well enough that even though the edge of totality moves at just over 1000 miles per hour, it's possible to predict when it will arrive at a given location to within perhaps a second. And as a demonstration of this, for the total eclipse back in 2017 we created a website to let anyone enter their geo location (or address) and then immediately compute when the eclipse would reach them—as well as generate many pages of other information.

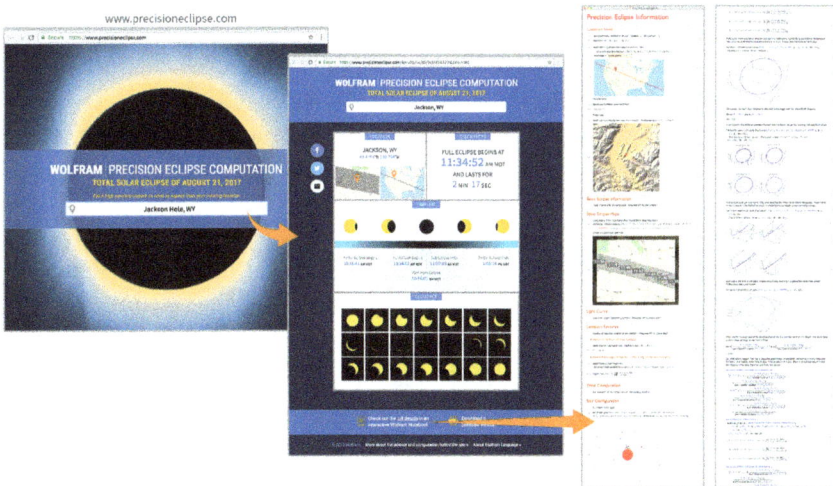

It's an Old Business

These days it's easy to find out when the next solar eclipse will be; indeed built right into the Wolfram Language there's just a function that tells you (in this form the output is the "time of greatest eclipse"):

In[ ]:= **SolarEclipse [ ]**

Out[ ]= Mon 8 Apr 2024 13:17:20 GMT−5

It's also easy to find out, and plot, where the region of totality will be:

In[ ]:= **GeoListPlot [SolarEclipse ["TotalPhasePolygon"]]**

Out[ ]=

Or to determine that the whole area of totality (including lots of ocean and some of Canada) will be about a third of the area of the US:

In[ ]:= **GeoArea [SolarEclipse ["TotalPhasePolygon"]] / GeoArea [🔲 USA]**

Out[ ]= 0.330287

But computing eclipses is not exactly a new business. In fact, the Antikythera device from 2000 years ago even tried to do it—using 37 metal gears to approximate the motion of the Sun and Moon (yes, with the Earth at the center). To me there's something unsettling—and cautionary—about the fact that the Antikythera device stands as such a solitary piece of technology, forgotten but not surpassed for more than 1600 years.

But right there on the bottom of the device there's an arm that moves around, and when it points to an H or Σ marking, it indicates a possible Sun or Moon eclipse. The way of setting dates on the device is a bit funky (after all, the modern calendar wouldn't be invented for another 1500 years), but if one takes the simulation on the Wolfram Demonstrations Project (which was calibrated back in 2012 when the Demonstration was created), and turns the crank to set the device for April 8, 2024, here's what one gets:

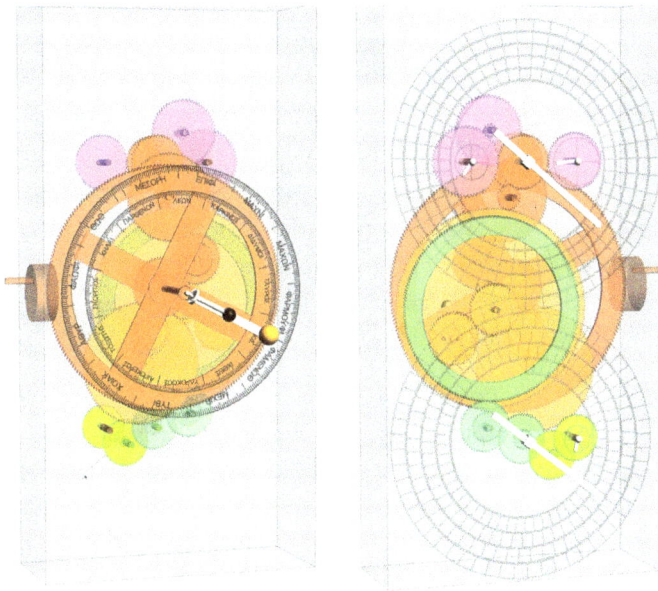

And, yes, all those gears move so as to line the Moon indicator up with the Sun— and to make the arm on the bottom point right at an H—just as it should for a solar eclipse. It's amazing to see this computation successfully happen on a device designed 2000 years ago.

Of course the results are a lot more accurate today. Though, strangely, despite all the theoretical science that's been done, the way we actually compute the position of the Sun and Moon is conceptually very much like the gears—and effectively epicycles—of the Antikythera device. It's just that now we have the digital equivalent of hundreds of thousands of gears.

Why Do Eclipses Happen?

A total solar eclipse occurs when the Moon gets in front of the Sun from the point of view of a particular location on the Earth. And it so happens that at this point in the Earth's history the Moon can just block the Sun because it has almost exactly the same angular diameter in the sky as the Sun (about 0.5° or 30 arc-minutes).

So when does the Moon get between the Sun and the Earth? Well, basically every time there's a new moon (i.e. once every lunar month). But we know there isn't an eclipse every month. So how come?

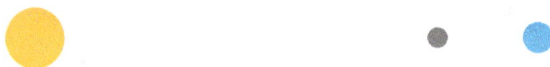

Well, actually, in the analogous situation of Ganymede and Jupiter, there *is* an eclipse every time Ganymede goes around Jupiter (which happens to be about once per week). Like the Earth, Jupiter's orbit around the Sun lies in a particular plane (the "plane of the ecliptic"). And it turns out that Ganymede's orbit around Jupiter also lies in essentially the same plane. So every time Ganymede reaches the "new moon" position (or, in official astronomy parlance, when it's aligned "in syzygy"—pronounced sizz-ee-gee), it's in the right place to cast its shadow onto Jupiter, and to eclipse the Sun wherever that shadow lands. (From Jupiter, Ganymede appears about 3 times the size of the Sun.)

But our Moon is different. Its orbit doesn't lie in the plane of the ecliptic. Instead, it's inclined at about 5°. (How it got that way is unknown, but it's presumably related to how the Moon was formed.) But that 5° is what makes eclipses so comparatively rare: they can only happen when there's a "new moon configuration" (syzygy) right at a time when the Moon's orbit passes through the plane of the ecliptic.

To show what's going on, let's draw an exaggerated version of everything. Here's the Moon going around the Earth, colored red whenever it's close to the plane of the ecliptic:

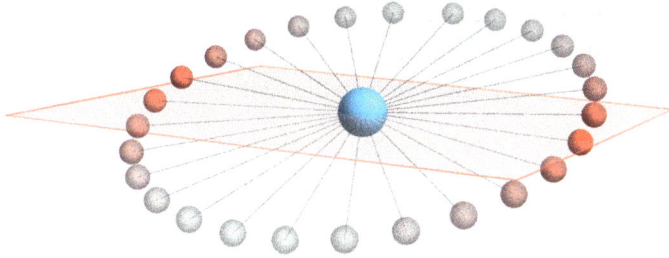

Now let's look at what happens over the course of about a year. We're showing a dot for where the Moon is each day. And the dot is redder if the Moon is closer to the plane of the ecliptic that day. (Note that if this was drawn to scale, you'd barely be able to see the Moon's orbit, and it wouldn't ever seem to go backwards like it does here.)

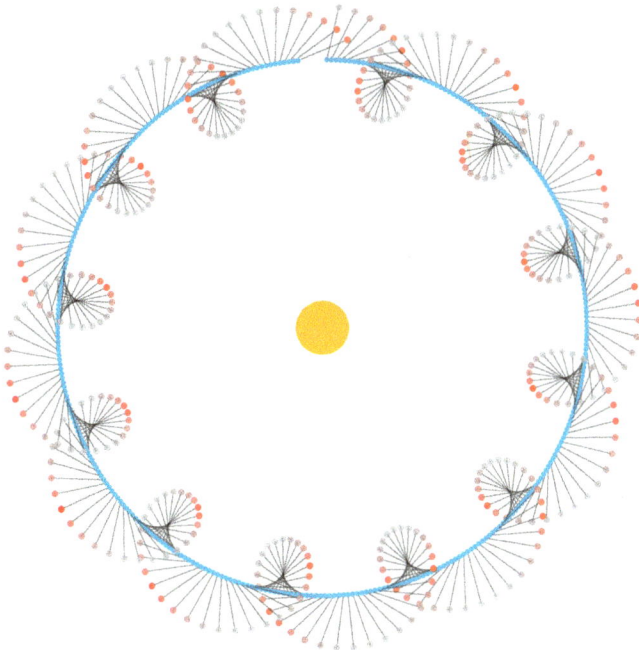

Now we can start to see how eclipses work. The basic point is that there's a solar eclipse whenever the Moon is both positioned between the Earth and the Sun, and it's in the plane of the ecliptic. In the picture, those two conditions correspond to the Moon being as far as possible towards the center, and as red as possible. So far we're only showing the position of the (exaggerated) Moon once per day. But to make things clearer, let's show it four times a day—and now prune out cases where the Moon isn't at least roughly lined up with the Sun:

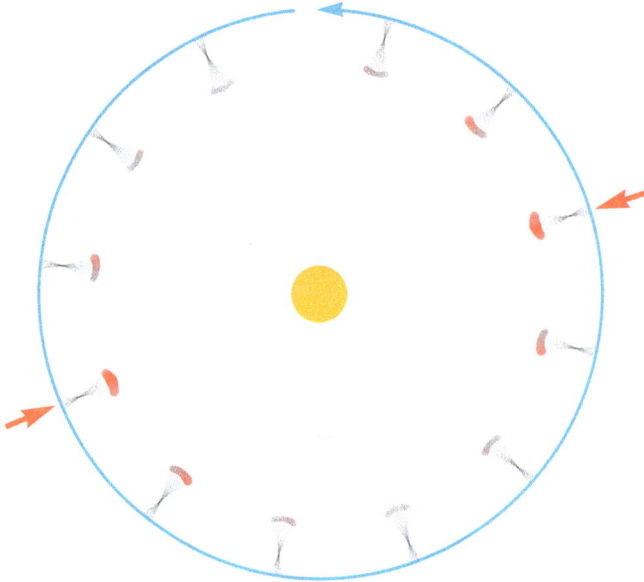

And now we can see that at least in this particular case, there are two points (indicated by arrows) where the Moon is lined up and in the plane of the ecliptic (so shown in red)—and these points will then correspond to solar eclipses.

In different years, the picture will look slightly different, essentially because the Moon is starting at a different place in its orbit at the beginning of the year. Here are schematic pictures for a few successive years:

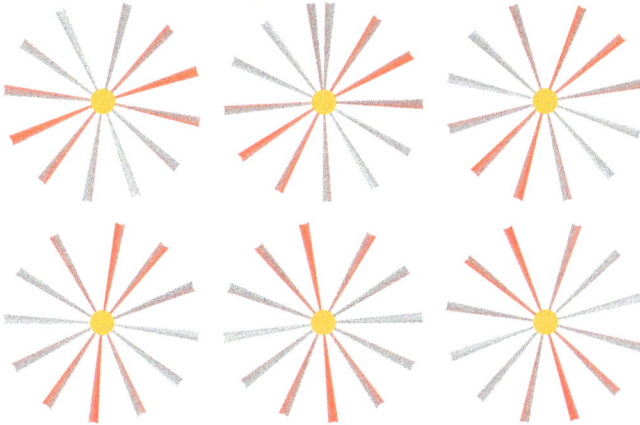

It's not so easy to see exactly when eclipses occur here—and it's also not possible to tell which are total eclipses where the Moon is exactly lined up, and which are only partial eclipses. But there's at least an indication, for example, that there are "eclipse seasons" in different parts of the year where eclipses happen.

OK, so what does the real data look like? Here's a plot for 20 years in the past and 20 years in the future, showing the actual days in each year when total and partial solar eclipses occur (the small dots everywhere indicate new moons):

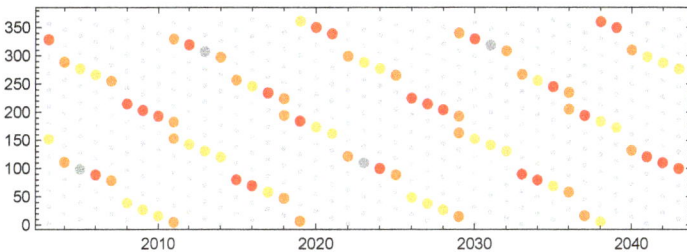

The reason for the "drift" between successive years is just that the lunar month (29.53 days) doesn't line up with the year, so the Moon doesn't go through a whole number of orbits in the course of a year, with the result that at the beginning of a new year, the Moon is in a different phase. But as the picture makes clear, there's quite a lot of regularity in the general times at which eclipses occur—and for example there are usually 2 eclipses in a given year—though there can be more (and in 0.2% of years there can be as many as 5, as there last were in 1935).

To see more detail about eclipses, let's plot the time differences (in fractional years) between all successive solar eclipses for 100 years in the past and 100 years in the future:

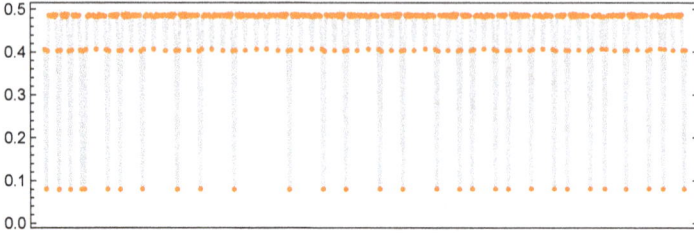

And now let's plot the same time differences, but just for total solar eclipses:

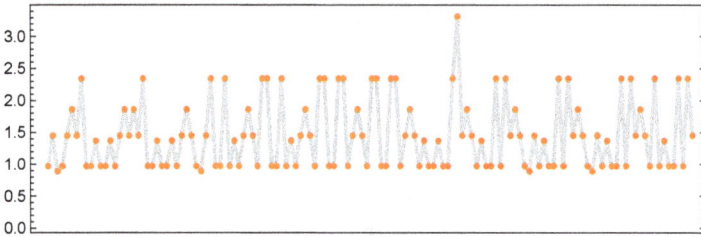

There's obviously a fair amount of overall regularity here, but there are also lots of little fine structure and irregularities. And being able to correctly predict all these details has basically taken science the better part of a few thousand years.

Ancient History

It's hard not to notice an eclipse, and presumably even from the earliest times people did. But were eclipses just reflections—or omens—associated with random goings-on in the heavens, perhaps in some kind of soap opera among the gods? Or were they things that could somehow be predicted?

A few thousand years ago, it wouldn't have been clear what people like astrologers could conceivably predict. When will the Moon be at a certain place in the sky? Will it rain tomorrow? What will the price of barley be? Who will win a battle? Even now, we're not sure how predictable all of these are. But the one clear case where prediction and exact science have triumphed is astronomy.

At least as far as the Western tradition is concerned, it all seems to have started in ancient Babylon—where for many hundreds of years, careful observations were made, and, in keeping with the ways of that civilization, detailed records were kept. And even today we still have thousands of daily official diary entries written in what look like tiny chicken scratches preserved on little clay tablets. "Night of the 14th: Cold north wind. Moon was in front of α Leonis. From 15th to 20th river rose $\frac{1}{2}$ cubit. Barley was 1 kur 5 siit. 25th, last part of night, moon was 1 cubit 8 fingers behind ϵ Leonis. 28th, 74° after sunrise, solar eclipse..."

If one looks at what happens on a particular day, one probably can't tell much. But by putting observations together over years or even hundreds of years, it's possible to see all sorts of repetitions and regularities. And back in Babylonian times the idea arose of using these to construct an ephemeris—a systematic table that said where a particular heavenly body such as the Moon was expected to be at any particular time.

(Needless to say, reconstructing Babylonian astronomy is a complicated exercise in decoding what's by now basically an alien culture. A key figure in this effort was a certain Otto Neugebauer, who happened to work down the hall from me at the Institute for Advanced Study in Princeton in the early 1980s. I would see him almost every day—a quiet white-haired chap, with a twinkle in his eye—and just sometimes I'd glimpse his huge filing system of index cards which I now realize was at the center of understanding Babylonian astronomy.)

One thing the Babylonians did was to measure surprisingly accurately the repetition period for the phases of the Moon—the so-called synodic month (or "lunation period") of about 29.53 days. And they noticed that 235 synodic months was very close to 19 years—so that about every 19 years, dates and phases of the Moon repeat their alignment, forming a so-called Metonic cycle (named after Meton of Athens, who described it in 432 BC).

It probably helps that the random constellations in the sky form a good pattern against which to measure the precise position of the Moon (it reminds me of the modern fashion of wearing fractals to make motion capture for movies easier). But the Babylonians noticed all sorts of details of the motion of the Moon. They knew about its "anomaly": its periodic speeding up and slowing down in the sky (now known to be a consequence of its slightly elliptical orbit). And they measured the average period of this—the so-called anomalistic month—to be about 27.55 days. They also noticed that the Moon went above and below the plane of the ecliptic (now known to be because of the inclination of its orbit)—with an average period (the so-called draconic month) that they measured as about 27.21 days.

And by 400 BC they'd noticed that every so-called saros of about 18 years 11 days all these different periods essentially line up (223 synodic months, 239 anomalistic months and 242 draconic months)—with the result that the Moon ends up at about the same position relative to the Sun. And this means that if there was an eclipse at one saros, then one can make the prediction that there's going to be an eclipse at the next saros too.

When one's absolutely precise about it, there are all sorts of effects that prevent precise repetition at each saros. But over timescales of more than 1300 years, there are in fact still strings of eclipses separated from each other by one saros. (Over the course of such a saros series, the locations of the eclipses effectively scan across the Earth; the upcoming eclipse is number 30 in a series of 71 that began in 1501 AD with an eclipse near the North Pole and will end in 2763 AD with an eclipse near the South Pole.)

Any given moment in time will be in the middle of quite a few saros series (right now it's 40)—and successive eclipses will always come from different series. But knowing about the saros cycle is a great first step in predicting eclipses—and it's for example what the Antikythera device uses. In a sense, it's a quintessential piece of science: take many observations, then synthesize a theory from them, or at least a scheme for computation.

It's not clear what the Babylonians thought about abstract, formal systems. But the Greeks were definitely into them. And by 300 BC Euclid had defined his abstract system for geometry. So when someone like Ptolemy did astronomy, they did it a bit like Euclid—effectively taking things like the saros cycle as axioms, and then proving from them often surprisingly elaborate geometrical theorems, such as that there must be at least two solar eclipses in a given year.

Ptolemy's *Almagest* from around 150 AD is an impressive piece of work, containing among many other things some quite elaborate procedures—and explicit tables— for predicting eclipses. (Yes, even in the later printed version, numbers are still represented confusingly by letters, as they always were in ancient Greek.)

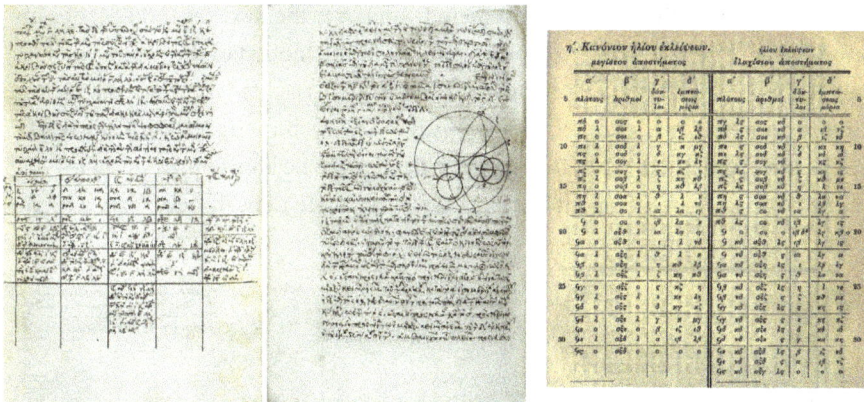

In Ptolemy's astronomy, Earth was assumed to be at the center of everything. But in modern terms that just meant he was choosing to use a different coordinate system—which didn't affect most of the things he wanted to do, like working out the geometry of eclipses. And unlike the mainline Greek philosophers he wasn't trying to make a fundamental theory of the world; he just wanted whatever epicycles and so on he needed to fit what he observed.

The Dawn of Modern Science

For more than a thousand years Ptolemy's theory of the Moon defined the state of the art. In the 1300s Ibn al-Shatir revised Ptolemy's models, achieving somewhat better accuracy. In 1472 Regiomontanus (Johannes Müller), systematizer of trigonometry, published more complete tables as part of his launch of what was essentially the first-ever scientific publishing company. But even in 1543 when Nicolaus Copernicus introduced his Sun-centered model of the solar system, the results he got were basically the same as Ptolemy's, even though his underlying description of what was going on was quite different.

It's said that Tycho Brahe got interested in astronomy in 1560 at age 13 when he saw a solar eclipse that had been predicted—and over the next several decades his careful observations uncovered several effects in the motion of the Moon (such as speeding up just before a full moon)—that eventually resulted in perhaps a factor 5 improvement in the prediction of its position. To Tycho eclipses were key tests, and he measured them carefully, and worked hard to be able to predict their timing more accurately than to within a few hours. (He himself never saw a total solar eclipse, only partial ones.)

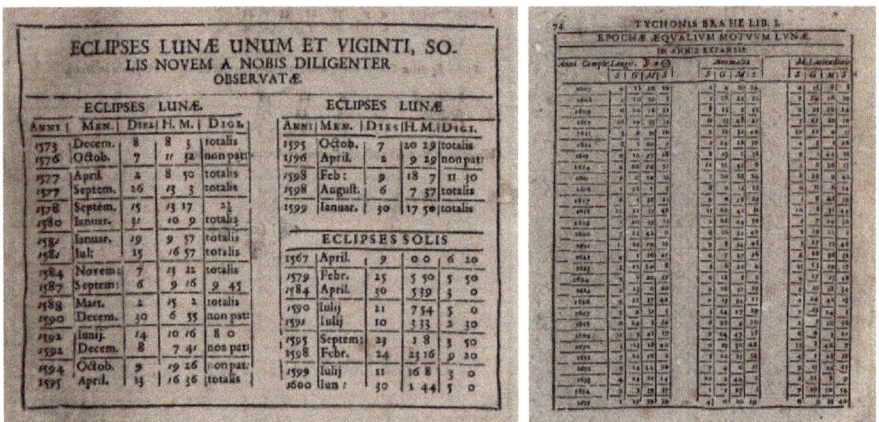

Armed with Tycho's observations, Johannes Kepler developed his description of orbits as ellipses—introducing concepts like inclination and eccentric anomaly—and in 1627 finally produced his *Rudolphine Tables*, which got right a lot of things that had been got wrong before, and included all sorts of detailed tables of lunar positions, as well as vastly better predictions for eclipses.

Using Kepler's *Rudolphine Tables* (and a couple of pages of calculations) the first known actual map of a solar eclipse was published in 1654. And while there are some charming inaccuracies in overall geography, the geometry of the eclipse isn't too bad.

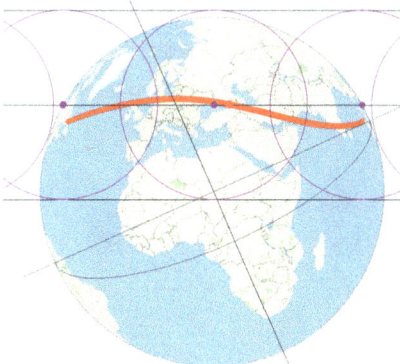

Whether it was Ptolemy's epicycles or Kepler's ellipses, there were plenty of calculations to do in determining the motions of heavenly bodies (and indeed the first known mechanical calculator—excepting the Antikythera device—was developed by a friend of Kepler's, presumably for the purpose). But there wasn't really a coherent underlying theory; it was more a matter of describing effects in ways that could be used to make predictions.

So it was a big step forward in 1687 when Isaac Newton published his *Principia*, and claimed that with his laws for motion and gravity it should be possible—essentially from first principles—to calculate everything about the motion of the Moon. (Charmingly, in his "Theory of the World" section he simply asserts as his Proposition XXII "That all the motions of the Moon… follow from the principles which we have laid down.")

Newton was proud of the fact that he could explain all sorts of known effects on the basis of his new theory. But when it came to actually calculating the detailed motion of the Moon he had a frustrating time. And even after several years he still couldn't get the right answer—in later editions of the *Principia* adding the admission that actually "The apse of the Moon is about twice as swift" (i.e. his answer was wrong by a factor of 2).

Still, in 1702 Newton was happy enough with his results that he allowed them to be published, in the form of a 20-page booklet on the "Theory of the Moon", which proclaimed that "By this Theory, what by all Astronomers was thought most difficult and almost impossible to be done, the Excellent Mr. Newton hath now effected, viz. to determine the Moon's Place even in her Quadratures, and all other Parts of her Orbit, besides the Syzygys, so accurately by Calculation, that the Difference between that and her true Place in the Heavens shall scarce be two Minutes…"

Newton didn't explain his methods (and actually it's still not clear exactly what he did, or how mathematically rigorous it was or wasn't). But his booklet effectively gave a step-by-step algorithm to compute the position of the Moon. He didn't claim it worked "at the syzygys" (i.e. when the Sun, Moon and Earth are lined up for an eclipse)—though his advertised error of two arc-minutes was still much smaller than the angular size of the Moon in the sky.

But it wasn't eclipses that were the focus then; it was a very practical problem of his day: knowing the location of a ship out in the open ocean. It's possible to determine what latitude you're at just by measuring how high the Sun gets in the sky. But to determine longitude you have to correct for the rotation of the Earth— and to do that you have to accurately keep track of time. But back in Newton's day, the clocks that existed simply weren't accurate enough, especially when they were being tossed around on a ship.

But particularly after various naval accidents, the problem of longitude was deemed important enough that the British government in 1714 established a "Board of Longitude" to offer prizes to help get it solved. One early suggestion was to use the regularity of the moons of Jupiter discovered by Galileo as a way to tell time. But it seemed that a simpler solution (not requiring a powerful telescope) might just be to measure the position of our Moon, say relative to certain fixed stars—and then to back-compute the time from this.

But to do this one had to have an accurate way to predict the motion of the Moon— which is what Newton was trying to provide. In reality, though, it took until the 1760s before tables were produced that were accurate enough to be able to determine time to within a minute (and thus distance to within 15 miles or so). And it so happens that right around the same time a marine chronometer was invented that was directly able to keep good time.

The Three-Body Problem

One of Newton's great achievements in the *Principia* was to solve the so-called two-body problem, and to show that with an inverse square law of gravity the orbit of one body around another must always be what Kepler had said: an ellipse.

In a first approximation, one can think of the Moon as just orbiting the Earth in a simple elliptical orbit. But what makes everything difficult is that that's just an approximation, because in reality there's also a gravitational pull on the Moon from the Sun. And because of this, the Moon's orbit is no longer a simple fixed ellipse—and in fact it ends up being much more complicated. There are a few definite effects one can describe and reason about. The ellipse gets stretched when the Earth is closer to the Sun in its own orbit. The orientation of the ellipse precesses like a top as a result of the influence of the Sun. But there's no way in the end to work out the orbit by pure reasoning—so there's no choice but to go into the mathematics and start solving the equations of the three-body problem.

In many ways this represented a new situation for science. In the past, one hadn't ever been able to go far without having to figure out new laws of nature. But here the underlying laws were supposedly known, courtesy of Newton. Yet even given these laws, there was difficult mathematics involved in working out the behavior they implied.

Over the course of the 1700s and 1800s the effort to try to solve the three-body problem and determine the orbit of the Moon was at the center of mathematical physics—and attracted a veritable who's who of mathematicians and physicists.

An early entrant was Leonhard Euler, who developed methods based on trigonometric series (including much of our current notation for such things), and whose works contain many immediately recognizable formulas:

In the mid-1740s there was a brief flap—also involving Euler's "competitors" Clairaut and d'Alembert—about the possibility that the inverse-square law for gravity might be wrong. But the problem turned out to be with the calculations, and by 1748 Euler was using sums of about 20 trigonometric terms and proudly proclaiming that the tables he'd produced for the three-body problem had predicted the time of a total solar eclipse to within minutes. (Actually, he had said there'd be 5 minutes of totality, whereas in reality there was only 1—but he blamed this error on incorrect coordinates he'd been given for Berlin.)

Mathematical physics moved rapidly over the next few decades, with all sorts of now-famous methods being developed, notably by people like Lagrange. And by the 1770s, for example, Lagrange's work was looking just like it could have come from a modern calculus book (or from a Wolfram|Alpha step-by-step solution):

Particularly in the hands of Laplace there was increasingly obvious success in deriving the observed phenomena of what he called "celestial mechanics" from mathematics—and in establishing the idea that mathematics alone could indeed generate new results in science.

At a practical level, measurements of things like the position of the Moon had always been much more accurate than calculations. But now they were becoming more comparable—driving advances in both. Meanwhile, there was increasing systematization in the production of ephemeris tables. And in 1767 the annual publication began of what was for many years the standard: the British *Nautical Almanac*.

The almanac quoted the position of the Moon to the arc-second, and systematically achieved at least arc-minute accuracy. The primary use of the almanac was for navigation (and it was what started the convention of using Greenwich as the "prime meridian" for measuring time). But right at the front of each year's edition were the predicted times of the eclipses for that year—in 1767 just two solar eclipses:

THE

NAUTICAL ALMANAC

AND

ASTRONOMICAL EPHEMERIS,

FOR THE YEAR 1767.

Published by ORDER of the

COMMISSIONERS OF LONGITUDE.

LONDON:

Printed by W. RICHARDSON and S. CLARK, PRINTERS;

AND SOLD BY

J. NOURSE, in the Strand, and Mess. MOUNT and PAGE, on Tower-Hill,

Booksellers to the said COMMISSIONERS. M DCC LXVI.

EXPLANATION of the Characters used in the EPHEMERIS.

The PLANETS, &c.

☉ The Sun.
☽ The Moon.
☿ Mercury.
♀ Venus.
☌ The Moon's, or any other Planet's Ascending Node.
☋ The Descending Node.
♂ Mars.
♃ Jupiter.
♄ Saturn.
☌ Conjunction, or Planets situated in the same Longitude.
☍ Opposition, or Planets situated in opposite Longitudes, or differing 6 Signs from each other.

Signs of the Zodiac.

♈ Aries.
♉ Taurus.
♊ Gemini.
♋ Cancer.
♌ Leo.
♍ Virgo.
♎ Libra.
♏ Scorpio.
♐ Sagittarius.
♑ Capricornus.
♒ Aquarius.
♓ Pisces.

ECLIPSES of the YEAR 1767.

Jan. 29. ☉ eclipsed, invisible in Europe. ☌ at 15ʰ. 43′. in 10°. 10″. 8′. with 1′. North Latitude.

July 25. ☉ eclipsed, begins at Sun-rising in Lat. S. 19°. 16′. Long. 141°. 45′. W. Ends at Sun-setting in Lat. 3°. 23′. S. Long. 50°. 5′. West. Centrally eclipsed on the Merid. in Lat. 1°. 15′. South.

	Obliquity of Ecliptic.	Equat. of Equinoct. Points.
Jan. 1.	23. 28. 17,2	+13,5
Apr. 1.	23. 28. 16,7	+14,5
July 1.	23. 28. 16,0	+15,5
Oct. 1.	23. 28. 15,2	+16,0
Dec. 31.	23. 28. 14,4	+16,6

JANUARY 1767.

Phases of the Moon.
First Quarter — 6. 20. 25
Full Moon — 14. 11. 40
Last Quarter — 22. 17. 39
New Moon — 29. 14. 43

The Math Gets More Serious

At a mathematical level, the three-body problem is about solving a system of three ordinary differential equations that give the positions of the three bodies as a function of time. If the positions are represented in standard 3D Cartesian coordinates $r_i = \{x_i, y_i, z_i\}$, the equations can be stated in the form:

$$m_{\mathbb{C}}\, \vec{r}_{\mathbb{C}}''(t) = -\, G\, m_{\mathbb{C}}\, m_{\oplus} \frac{\vec{r}_{\mathbb{C}}(t) - \vec{r}_{\oplus}(t)}{|\vec{r}_{\mathbb{C}}(t) - \vec{r}_{\oplus}(t)|^{3/2}} - G\, m_{\mathbb{C}}\, m_{\odot} \frac{\vec{r}_{\mathbb{C}}(t) - \vec{r}_{\odot}(t)}{|\vec{r}_{\mathbb{C}}(t) - \vec{r}_{\odot}(t)|^{3/2}}$$

$$m_{\oplus}\, \vec{r}_{\oplus}''(t) = -\, G\, m_{\oplus}\, m_{\mathbb{C}} \frac{\vec{r}_{\oplus}(t) - \vec{r}_{\mathbb{C}}(t)}{|\vec{r}_{\oplus}(t) - \vec{r}_{\mathbb{C}}(t)|^{3/2}} - G\, m_{\oplus}\, m_{\odot} \frac{\vec{r}_{\oplus}(t) - \vec{r}_{\odot}(t)}{|\vec{r}_{\oplus}(t) - \vec{r}_{\odot}(t)|^{3/2}}$$

$$m_{\odot}\, \vec{r}_{\odot}''(t) = -\, G\, m_{\odot}\, m_{\oplus} \frac{\vec{r}_{\odot}(t) - \vec{r}_{\oplus}(t)}{|\vec{r}_{\odot}(t) - \vec{r}_{\oplus}(t)|^{3/2}} - G\, m_{\odot}\, m_{\mathbb{C}} \frac{\vec{r}_{\odot}(t) - \vec{r}_{\mathbb{C}}(t)}{|\vec{r}_{\odot}(t) - \vec{r}_{\mathbb{C}}(t)|^{3/2}}$$

The $\{x, y, z\}$ coordinates here aren't, however, what traditionally show up in astronomy. For example, in describing the position of the Moon one might use longitude and latitude on a sphere around the Earth. Or, given that one knows the Moon has a roughly elliptical orbit, one might instead choose to describe its motions by variables that are based on deviations from such an orbit. In principle it's just a matter of algebraic manipulation to restate the equations with any given choice of variables. But in practice what comes out is often long and complex—and can lead to formulas that fill many pages.

But, OK, so what are the best kinds of variables to use for the three-body problem? Maybe they should involve relative positions of pairs of bodies. Or relative angles. Or maybe positions in various kinds of rotating coordinate systems. Or maybe quantities that would be constant in a pure two-body problem. Over the course of the 1700s and 1800s many treatises were written exploring different possibilities.

But in essentially all cases the ultimate approach to the three-body problem was the same. Set up the problem with the chosen variables. Identify parameters that, if set to zero, would make the problem collapse to some easy-to-solve form. Then do a series expansion in powers of these parameters, keeping just some number of terms.

By the 1860s Charles Delaunay had spent 20 years developing the most extensive theory of the Moon in this way. He'd identified five parameters with respect to which to do his series expansions (eccentricities, inclinations, and ratios of orbit sizes)—and in the end he generated about 1800 pages like this (yes, he really needed Mathematica!):

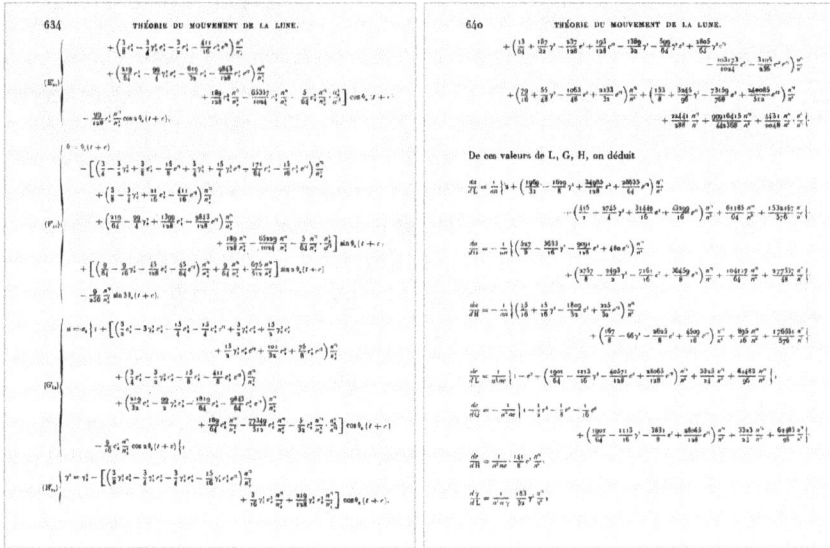

But the sad fact was that despite all this effort, he didn't get terribly good answers. And eventually it became clear why. The basic problem was that Delaunay wanted to represent his results in terms of functions like sin and cos. But in his computations, he often wanted to do series expansions with respect to the frequencies of those functions. Here's a minimal case:

In[ ]:= **Series [Sin [(ω + δ) t] , { δ, 0, 3 }]**

Out[ ]= $\text{Sin}[t\omega] + t\text{Cos}[t\omega]\,\delta - \frac{1}{2}(t^2\text{Sin}[t\omega])\,\delta^2 - \frac{1}{6}(t^3\text{Cos}[t\omega])\,\delta^3 + O[\delta]^4$

And here's the problem. Take a look even at the second term. Yes, the δ parameter may be small. But how about the t parameter, standing for time? If you don't want to make predictions very far out, that'll stay small. But what if you want to figure out what will happen further in the future?

Well, eventually that term will get big. And higher-order terms will get even bigger. But unless the Moon is going to escape its orbit or something, the final mathematical expressions that represent its position can't have values that are too big. So in these expressions the so-called secular terms that increase with t must somehow cancel out.

21

But the problem is that at any given order in the series expansion, there's no guarantee that will happen in a numerically useful way. And in Delaunay's case— even though with immense effort he often went to 7th order or beyond—it didn't.

One nice feature of Delaunay's computation was that it was in a sense entirely algebraic: everything was done symbolically, and only at the very end were actual numerical values of parameters substituted in.

But even before Delaunay, Peter Hansen had taken a different approach— substituting numbers as soon as he could, and dropping terms based on their numerical size rather than their symbolic form. His presentations look less pure (notice things like all those t–1800, where t is the time in years), and it's more difficult to tell what's going on. But as a practical matter, his results were much better, and in fact were used for many national almanacs from about 1862 to 1922, achieving errors as small as 1 or 2 arc-seconds at least over periods of a decade or so. (Over longer periods, the errors could rapidly increase because of the lack of terms that had been dropped as a result of what amounted to numerical accidents.)

Both Delaunay and Hansen tried to represent orbits as series of powers and trigonometric functions (so-called Poisson series). But in the 1870s, George Hill in the US Nautical Almanac Office proposed instead using as a basis numerically computed functions that came from solving an equation for two-body motion with a periodic driving force of roughly the kind the Sun exerts on the Moon's orbit. A large-scale effort was mounted, and starting in 1892 Ernest W. Brown (who had moved to the US, but had been a student of George Darwin, Charles Darwin's physicist son) took charge of the project and in 1918 produced what would stand for many years as the definitive "Tables of the Motion of the Moon".

Brown's tables consist of hundreds of pages like this—ultimately representing the position of the Moon as a combination of about 1400 terms with very precise coefficients:

TERMS IN THE TABLES OF SECT. III	63	64	TABLES OF THE MOON, SECT. I, CHAP. IV.

He says right at the beginning that the tables aren't particularly intended for unique events like eclipses, but then goes ahead and does a "worked example" of computing an eclipse from 381 BC, reported by Ptolemy:

It was an impressive indication of how far things had come. But ironically enough the final presentation of Brown's tables had the same sum-of-trigonometric-functions form that one would get from having lots of epicycles. At some level it's not surprising, because any function can ultimately be represented by epicycles, just as it can be represented by a Fourier or other series. But it's a strange quirk of history that such similar forms were used.

Can the Three-Body Problem Be Solved?

It's all well and good that one can find approximations to the three-body problem, but what about just finding an outright solution—like as a mathematical formula? Even in the 1700s, there'd been some specific solutions found—like Euler's collinear configuration, and Lagrange's equilateral triangle. But a century later, no further solutions had been found—and finding a complete solution to the three-body problem was beginning to seem as hopeless as trisecting an angle, solving the quintic, or making a perpetual motion machine. (That sentiment was reflected for example in a letter Charles Babbage wrote Ada Lovelace in 1843 mentioning the "horrible problem [of] the three bodies"—even though this letter was later misinterpreted by Ada's biographers to be about a romantic triangle, not the three-body problem of celestial mechanics.)

In contrast to the three-body problem, what seemed to make the two-body problem tractable was that its solutions could be completely characterized by "constants of the motion"—quantities that stay constant with time (in this case notably the direction of the axis of the ellipse). So for many years one of the big goals with the three-body problem was to find constants of the motion.

In 1887, though, Heinrich Bruns showed that there couldn't be any such constants of the motion, at least expressible as algebraic functions of the standard $\{x, y, z\}$ position and velocity coordinates of the three bodies. Then in the mid-1890s Henri Poincaré showed that actually there couldn't be any constants of the motion that were expressible as any analytic functions of the positions, velocities and mass ratios.

One reason that was particularly disappointing at the time was that it had been hoped that somehow constants of the motion would be found in n-body problems that would lead to a mathematical proof of the long-term stability of the solar system. And as part of his work, Poincaré also saw something else: that at least in particular cases of the three-body problem, there was arbitrarily sensitive dependence on initial conditions—implying that even tiny errors in measurement could be amplified to arbitrarily large changes in predicted behavior (the classic "chaos theory" phenomenon).

But having discovered that particular solutions to the three-body problem could have this kind of instability, Poincaré took a different approach that would actually be characteristic of much of pure mathematics going forward: he decided to look not at individual solutions, but at the space of all possible solutions. And needless to say, he found that for the three-body problem, this was very complicated—though in his efforts to analyze it he invented the field of topology.

Poincaré's work all but ended efforts to find complete solutions to the three-body problem. It also seemed to some to explain why the series expansions of Delaunay and others hadn't worked out—though in 1912 Karl Sundman did show that at least in principle the three-body problem could be solved in terms of an infinite series, albeit one that converges outrageously slowly.

But what does it mean to say that there can't be a solution to the three-body problem? Galois had shown that there couldn't be a solution to the generic quintic equation, at least in terms of radicals. But actually it's still perfectly possible to express the solution in terms of elliptic or hypergeometric functions. So why can't there be some more sophisticated class of functions that can be used to just "solve the three-body problem"?

Here are some pictures of what can actually happen in the three-body problem, with various initial conditions:

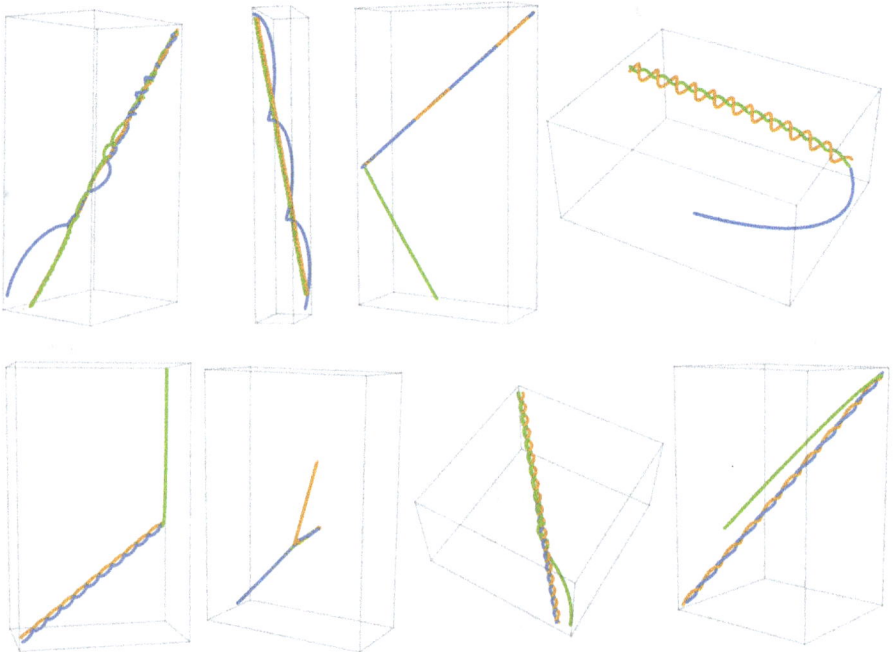

And looking at these immediately gives some indication of why it's not easy to just "solve the three-body problem". Yes, there are cases where what happens is fairly simple. But there are also cases where it's not, and where the trajectories of the three bodies continue to be complicated and tangled for a long time.

So what's fundamentally going on here? I don't think traditional mathematics is the place to look. But I think what we're seeing is actually an example of a general phenomenon I call computational irreducibility that I discovered in the 1980s in studying the computational universe of possible programs.

Many programs, like many instances of the three-body problem, behave in quite simple ways. But if you just start looking at all possible simple programs, it doesn't take long before you start seeing behavior like this:

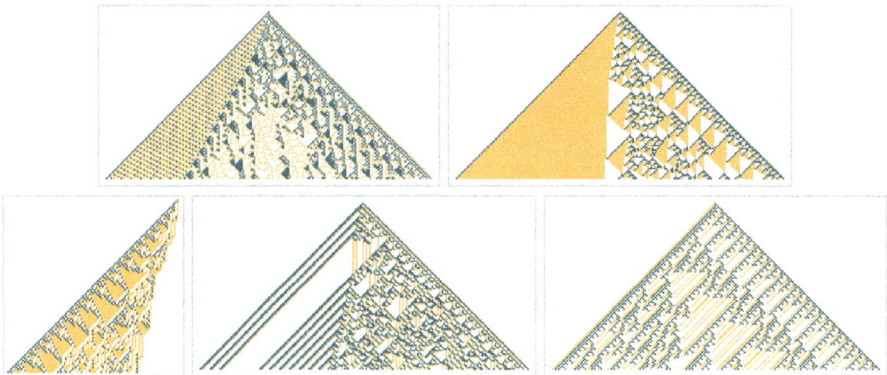

How can one tell what's going to happen? Well, one can just keep explicitly running each program and seeing what it does. But the question is: is there some systematic way to jump ahead, and to predict what will happen without tracing through all the steps?

The answer is that in general there isn't. And what I call the Principle of Computational Equivalence suggests that pretty much whenever one sees complex behavior, there won't be.

Here's the way to think about this. The system one's studying is effectively doing a computation to work out what its behavior will be. So to jump ahead we'd in a sense have to do a more sophisticated computation. But what the Principle of Computational Equivalence says is that actually we can't—and that whether we're using our brains or our mathematics or a Turing machine or anything else, we're always stuck with computations of the same sophistication.

So what about the three-body problem? Well, I strongly suspect that it's an example of computational irreducibility: that in effect the computations it's doing are as sophisticated as any computations that we can do, so there's no way we can ever expect to systematically jump ahead and solve the problem. (We also can't expect to just define some new finite class of functions that can just be evaluated to give the solution.)

I'm hoping that one day someone will rigorously prove this. There's some technical difficulty, because the three-body problem is usually formulated in terms of real numbers that immediately have an infinite number of digits—but to compare with ordinary computation one has to require finite processes to set up initial conditions. (Ultimately one wants to show for example that there's a "compiler" that can go from any program, say for a Turing machine, and can generate instructions to set up initial conditions for a three-body problem so that the evolution of the three-body problem will give the same results as running that program—implying that the three-body problem is capable of universal computation.)

I have to say that I consider Newton in a sense very lucky. It could have been that it wouldn't have been possible to work out anything interesting from his theory without encountering the kind of difficulties he had with the motion of the Moon—because one would always be running into computational irreducibility. But in fact, there was enough computational reducibility and enough that could be computed easily that one could see that the theory was useful in predicting features of the world (and not getting wrong answers, like with the apse of the Moon)—even if there were some parts that might take two centuries to work out, or never be possible at all.

Newton himself was certainly aware of the potential issue, saying that at least if one was dealing with gravitational interactions between many planets then "to define these motions by exact laws admitting of easy calculation exceeds, if I am not mistaken, the force of any human mind". And even today it's extremely difficult to know what the long-term evolution of the solar system will be.

It's not particularly that there's sensitive dependence on initial conditions: we actually have measurements that should be precise enough to determine what will happen for a long time. The problem is that we just have to do the computation—a bit like computing the digits of π—to work out the behavior of the n-body problem that is our solar system.

Existing simulations show that for perhaps a few tens of millions of years, nothing too dramatic can happen. But after that we don't know. Planets could change their order. Maybe they could even collide, or be ejected from the solar system. Computational irreducibility implies that at least after an infinite time it's actually formally undecidable (in the sense of Gödel's theorem or the halting problem) what can happen.

One of my children, when they were very young, asked me whether when dinosaurs existed the Earth could have had two moons. For years when I ran into celestial mechanics experts I would ask them that question—and it was notable how difficult they found it. Most now say that at least at the time of the dinosaurs we couldn't have had an extra moon—though a billion years earlier it's not clear.

We used to only have one system of planets to study. And the fact that there were (then) 9 of them used to be a classic philosopher's example of a truth about the world that just happens to be the way it is, and isn't "necessarily true" (like $2 + 2 = 4$). But now of course we know about lots of exoplanets. And it's beginning to look as if there might be a theory for things like how many planets a solar system is likely to have.

At some level there's presumably a process like natural selection: some configurations of planets aren't "fit enough" to be stable—and only those that are survive. In biology it's traditionally been assumed that natural selection and adaptation is somehow what's led to the complexity we see. But actually I suspect much of it is instead just a reflection of what generally happens in the computational universe—both in biology and in celestial mechanics. Now in celestial mechanics, we haven't yet seen in the wild any particularly complex forms (beyond a few complicated gap structures in rings, and tumbling moons and asteroids). But perhaps elsewhere we'll see things like those obviously tangled solutions to the three-body problem—that come closer to what we're used to in biology.

It's remarkable how similar the issues are across so many different fields. For example, the whole idea of using "perturbation theory" and series expansions that has existed since the 1700s in celestial mechanics is now also core to quantum field theory. But just like in celestial mechanics there's trouble with convergence (maybe one should try renormalization or resummation in celestial mechanics). And in the end one begins to realize that there are phenomena—no doubt like turbulence or the three-body problem—that inevitably involve more sophisticated computations, and that need to be studied not with traditional mathematics of the kind that was so successful for Newton and his followers but with the kind of science that comes from exploring the computational universe.

Approaching Modern Times

But let's get back to the story of the motion of the Moon. Between Brown's tables, and Poincaré's theoretical work, by the beginning of the 1900s the general impression was that whatever could reasonably be computed about the motion of the Moon had been computed.

Occasionally there were tests. Like for example in 1925, when there was a total solar eclipse visible in New York City, and the *New York Times* perhaps overdramatically said that "scientists [were] tense... [wondering] whether they or Moon is wrong as eclipse lags five seconds behind". The fact is that a prediction accurate to 5 seconds was remarkably good, and we can't do all that much better even today. (By the way, the actual article talks extensively about "Professor Brown"— as well as about how the eclipse might "disprove Einstein" and corroborate the existence of "coronium"—but doesn't elaborate on the supposed prediction error.)

As a practical matter, Brown's tables were not exactly easy to use: to find the position of the Moon from them required lots of mechanical desk calculator work, as well as careful transcription of numbers. And this led Leslie Comrie in 1932 to propose using a punch-card-based IBM Hollerith automatic tabulator—and with the help of Thomas Watson, CEO of IBM, what was probably the first "scientific computing laboratory" was established—to automate computations from Brown's tables.

© Royal Astronomical Society

(When I was in elementary school in England in the late 1960s—before electronic calculators—I always carried around, along with my slide rule, a little book of "4-figure mathematical tables". I think I found it odd that such a book would have an author—and perhaps for that reason I still remember the name: "L. J. Comrie".)

By the 1950s, the calculations in Brown's tables were slowly being rearranged and improved to make them more suitable for computers. But then with John F. Kennedy's 1962 "We choose to go the Moon", there was suddenly urgent interest in getting the most accurate computations of the Moon's position. As it turned out, though, it was basically just a tweaked version of Brown's tables, running on a mainframe computer, that did the computations for the Apollo program.

At first, computers were used in celestial mechanics purely for numerical computa-tion. But by the mid-1960s there were also experiments in using them for algebraic computation, and particularly to automate the generation of series expansions. Wallace Eckert at IBM started using FORMAC to redo Brown's tables, while in Cambridge David Barton and Steve Bourne (later the creator of the "Bourne shell" (sh) in Unix) built their own CAMAL computer algebra system to try extending the kind of thing Delaunay had done. (And by 1970, Delaunay's 7th-order calculations had been extended to 20th order.)

When I myself started to work on computer algebra in 1976 (primarily for computations in particle physics), I'd certainly heard about CAMAL, but I didn't know what it had been used for (beyond vaguely "celestial mechanics"). And as a practicing theoretical physicist in the late 1970s, I have to say that the "problem of the Moon" that had been so prominent in the 1700s and 1800s had by then fallen into complete obscurity.

I remember for example in 1984 asking a certain Martin Gutzwiller, who was talking about quantum chaos, what his main interest actually was. And when he said "the problem of the Moon", I was floored; I didn't know there still was any problem with the Moon. As it turns out, in writing this post I found out that Gutzwiller was actually the person who took over from Eckert and spent nearly two decades working on trying to improve the computations of the position of the Moon.

Why Not Just Solve It?

Traditional approaches to the three-body problem come very much from a mathematical way of thinking. But modern computational thinking immediately suggests a different approach. Given the differential equations for the three-body problem, why not just directly solve them? And indeed in the Wolfram Language there's a built-in function NDSolve for numerically solving systems of differential equations.

So what happens if one just feeds in equations for a three-body problem? Well, here are the equations:

$In[\circ]:=$ eqns $= \Big\{ m_1 r_1''[t] == -\dfrac{m_1 m_2 (r_1[t] - r_2[t])}{\text{Norm}[r_1[t] - r_2[t]]^3} - \dfrac{m_1 m_3 (r_1[t] - r_3[t])}{\text{Norm}[r_1[t] - r_3[t]]^3},$

$m_2 r_2''[t] == -\dfrac{m_1 m_2 (r_2[t] - r_1[t])}{\text{Norm}[r_2[t] - r_1[t]]^3} - \dfrac{m_2 m_3 (r_2[t] - r_3[t])}{\text{Norm}[r_2[t] - r_3[t]]^3},$

$m_3 r_3''[t] == -\dfrac{m_1 m_3 (r_3[t] - r_1[t])}{\text{Norm}[r_3[t] - r_1[t]]^3} - \dfrac{m_2 m_3 (r_3[t] - r_2[t])}{\text{Norm}[r_3[t] - r_2[t]]^3} \Big\};$

Now as an example let's set the masses to random values:

$In[\circ]:=$ $\{m_1, m_2, m_3\}$ = RandomReal[{0, 1}, 3]

$Out[\circ]=$ {0.43169, 0.60114, 0.46066}

And let's define the initial position and velocity for each body to be random as well:

$In[\circ]:=$ inits = Table[{ r_i[0] == RandomReal[{-1, 1}, 3],
r_i'[0] == RandomReal[{-1, 1}, 3] }, {i, 3}]

$Out[\circ]=$ { { r_1[0] == {-0.275557, 0.715307, 0.390286}, r_1'[0] == {-0.156962, 0.977674, 0.424888} },
{ r_2[0] == {0.992591, 0.138954, -0.863011}, r_2'[0] == {-0.140601, 0.985443, 0.196193} },
{ r_3[0] == {0.900842, 0.243414, -0.514439}, r_3'[0] == {0.570242, -0.664676, -0.528499} } }

Now we can just use NDSolve to get the solutions (it gives them as implicit approximate numerical functions of t):

$In[\circ]:=$ sols = NDSolve[{ eqns, inits }, {r_1, r_2, r_3}, {t, 0, 100}]

$Out[\circ]=$ { { $r_1 \to$ InterpolatingFunction[Domain: {{0., 100. }} Output dimensions: {3}],

$r_2 \to$ InterpolatingFunction[Domain: {{0., 100. }} Output dimensions: {3}],

$r_3 \to$ InterpolatingFunction[Domain: {{0., 100. }} Output dimensions: {3}] } }

And now we can plot them. And now we've got a solution to a three-body problem, just like that!

In[◦]:= **ParametricPlot3D [Evaluate [{ r_1 [t] , r_2 [t] , r_3 [t] } /. First [sols]] , { t, 0, 100 }]**

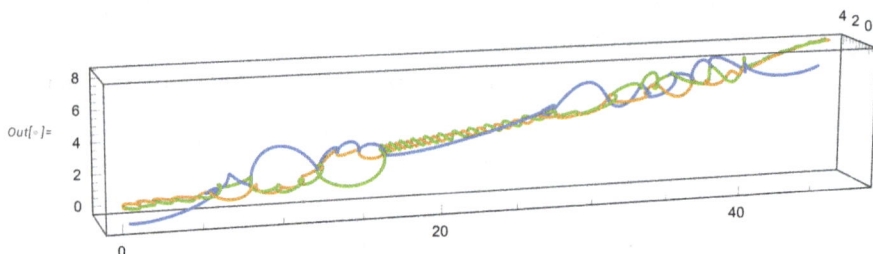

Well, obviously this is using the Wolfram Language and a huge tower of modern technology. But would it have been possible even right from the beginning for people to generate direct numerical solutions to the three-body problem, rather than doing all that algebra? Back in the 1700s, Euler already knew what's now called Euler's method for finding approximate numerical solutions to differential equations. So what if he'd just used that method to calculate the motion of the Moon?

The method relies on taking a sequence of discrete steps in time. And if he'd used, say, a step size of a minute, then he'd have had to take 40,000 steps to get results for a month, but he should have been able to successfully reproduce the position of the Moon to about a percent. If he'd tried to extend to 3 months, however, then he would already have had at least a 10% error.

Any numerical scheme for solving differential equations in practice eventually builds up some kind of error—but the more one knows about the equations one's solving, and their expected solutions, the more one's able to preprocess and adapt things to minimize the error. NDSolve has enough automatic adaptivity built into it that it'll do pretty well for a surprisingly long time on a typical three-body problem. (It helps that the Wolfram Language and NDSolve can handle numbers with arbitrary precision, not just machine precision.)

But if one looks, say, at the total energy of the three-body system—which one can prove from the equations should stay constant—then one will typically see an error slowly build up in it. One can avoid this if one effectively does a change of variables in the equations to "factor out" energy. And one can imagine doing a whole hierarchy of algebraic transformations that in a sense give the numerical scheme as much help as possible.

And indeed since at least the 1980s that's exactly what's been done in practical work on the three-body problem, and the Earth-Moon-Sun system. So in effect it's a mixture of the traditional algebraic approach from the 1700s and 1800s, together with modern numerical computation.

The Real Earth-Moon-Sun Problem

OK, so what's involved in solving the real problem of the Earth-Moon-Sun system? The standard three-body problem gives a remarkably good approximation to the physics of what's happening. But it's obviously not the whole story.

For a start, the Earth isn't the only planet in the solar system. And if one's trying to get sufficiently accurate answers, one's going to have to take into account the gravitational effect of other planets. The most important is Jupiter, and its typical effect on the orbit of the Moon is at about the 10^{-5} level—sufficiently large that for example Brown had to take it into account in his tables.

The next effect is that the Earth isn't just a point mass, or even a precise sphere. Its rotation makes it bulge at the equator, and that affects the orbit of the Moon at the 10^{-6} level.

Orbits around the Earth ultimately depend on the full mass distribution and gravitational field of the Earth (which is what Sputnik-1 was nominally launched to map)—and both this, and the reverse effect from the Moon, come in at the 10^{-8} level. At the 10^{-9} level there are then effects from tidal deformations ("solid tides") on the Earth and Moon, as well as from gravitational redshift and other general relativistic phenomena.

To predict the position of the Moon as accurately as possible one ultimately has to have at least some model for these various effects.

But there's a much more immediate issue to deal with: one has to know the initial conditions for the Earth, Sun and Moon, or in other words, one has to know as accurately as possible what their positions and velocities were at some particular time.

And conveniently enough, there's now a really good way to do that, because Apollo 11, 14 and 15 all left laser retroreflectors on the Moon. And by precisely timing how long it takes a laser pulse from the Earth to round-trip to these retroreflectors, it's now possible in effect to measure the position of the Moon to millimeter accuracy.

OK, so how do modern analogs of the Babylonian ephemerides actually work? Internally they're dealing with the equations for all the significant bodies in the solar system. They do symbolic preprocessing to make their numerical work as easy as possible. And then they directly solve the differential equations for the system, appropriately inserting models for things like the mass distribution in the Earth.

They start from particular measured initial conditions, but then they repeatedly insert new measurements, trying to correct the parameters of the model so as to optimally reproduce all the measurements they have. It's very much like a typical machine learning task—with the training data here being observations of the solar system (and typically fitting just being least squares).

But, OK, so there's a model one can run to figure out something like the position of the Moon. But one doesn't want to have to explicitly do that every time one needs to get a result; instead one wants in effect just to store a big table of pre-computed results, and then to do something like interpolation to get any particular result one needs. And indeed that's how it's done today.

How It's Really Done

Back in the 1960s NASA started directly solving differential equations for the motion of planets. The Moon was more difficult to deal with, but by the 1980s that too was being handled in a similar way. Ongoing data from things like the lunar retroreflectors was added, and all available historical data was inserted as well.

The result of all this was the JPL Development Ephemeris (JPL DE). In addition to new observations being used, the underlying system gets updated every few years, for example to get what's needed for some spacecraft going to some new place in the solar system. (The latest is DE441—that follows DE432, which was built for going to Pluto.)

But so how is the actual ephemeris delivered? Well, for every thousand years covered, the ephemeris has about 100 megabytes of results, given as coefficients for Chebyshev polynomials, which are convenient for interpolation. And for any given quantity in any given coordinate system over a particular period of time, one accesses the appropriate parts of these results.

In Wolfram Language, it's all packaged up into the function AstroPosition—which here gives the position of the Moon in coordinates relative to the equator of the Earth right now:

In[ ]:= **AstroPosition ["Moon", "Equatorial"]**

Out[ ]:= AstroPosition[Right ascension: $4^h 40^m 54.3^s$ Declination: $26°3'36.5"$]

OK, but so how does one find an eclipse? Well, it's an iterative process. Start with an approximation, perhaps from the saros cycle. Then interpolate the ephemeris and look at the result. Then keep iterating until one finds out just when the Moon will be in the appropriate position.

But actually there's some more to do. Because what's originally computed are the positions of the barycenters (centers of mass) of the various bodies. But now one has to figure out how the bodies are oriented.

The Earth rotates, and we know its rate quite precisely. But the Moon is basically locked with the same face pointing to the Earth, except that in practice there are small "librations" where the Moon wobbles a little back and forth—and these turn out to be particularly troublesome to predict.

Where Will the Eclipse Be?

OK, so let's say one knows where the Earth, Moon and Sun are. How does one then figure out what type of eclipse will happen, and where on the Earth the eclipse will actually hit? Well, there's some further geometry to do. And here's the beginning of what's involved:

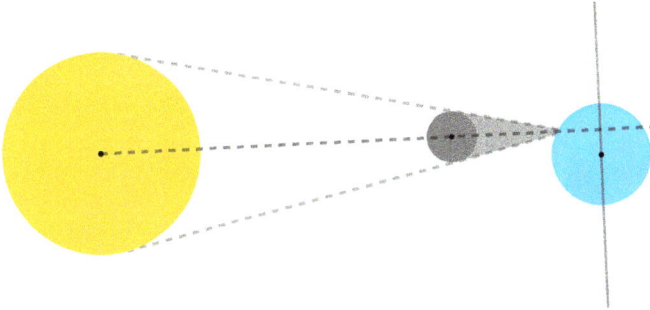

Basically the Moon generates a cone of shadow, and then the question is how this cone intersects the Earth. If the tip of the cone is inside the Earth, that means there'll be a region of total shadow ("umbra") on the Earth—and a total eclipse. (If the tip is above the Earth, there'll be an annular eclipse, in which there's a "ring of sun" visible around the Moon.)

By the way, the more complete geometry is like this (again not to scale)

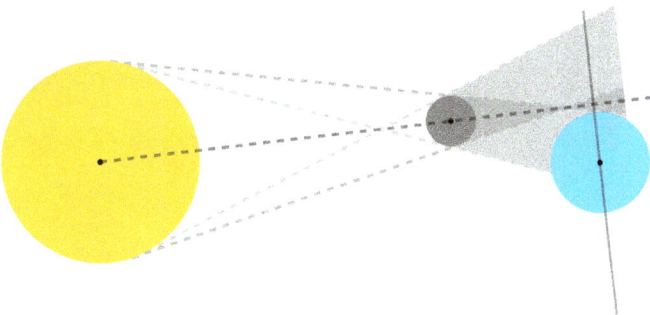

where now we've included the penumbra in which only part of the Sun is shadowed by the Moon. In the particular case shown, the umbra cone "misses the Earth", so there's no total eclipse, but there's still a partial eclipse where part of the Sun is shadowed.

OK, but let's say there's going to be a total eclipse. To see where on the Earth the region of totality will be, we have to work out where the "cone of total shadow" (i.e. umbra) will intersect the surface of the Earth. It's a somewhat complicated 3D geometry problem:

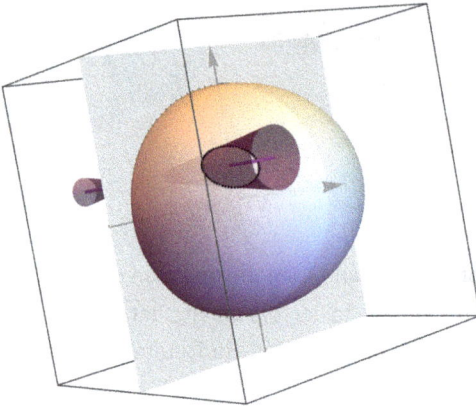

It's easiest to understand what happens by looking at things from the position of the Sun. The light gray region is the penumbra, and the little black dot is the region of totality (i.e. the umbra):

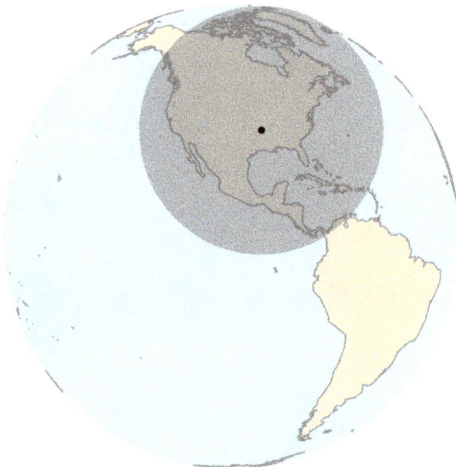

As the Earth and Moon move in their orbits, the region of shadow will move relative to the Earth:

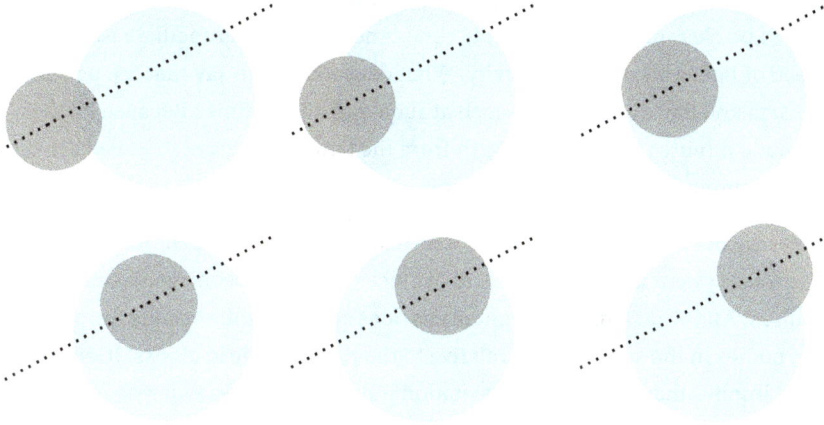

But now there's another part of the story, which is the rotation of the Earth. And if we include that, we'll see that the region of totality (at least in this case) traces out a kind of S-shaped curve on the surface of the Earth:

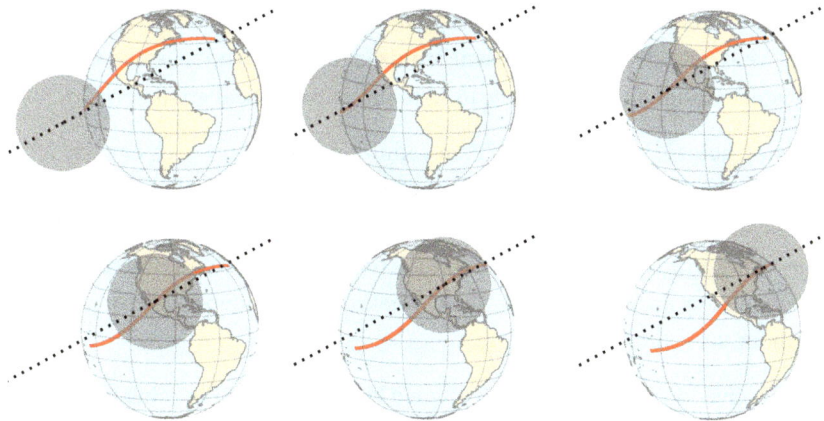

All this tricky geometry got figured out in 1824 by Friedrich Bessel, who introduced what are now called the Besselian elements—eight variables that specify the location, orientation and aperture of the umbra and penumbra cones, as well as the orientation of the Earth, as a function of time. And for any given eclipse, the whole story of its appearance and trajectory on the surface of the Earth is determined by its Besselian elements.

When Will the Eclipse Arrive?

OK, so now we know what the trajectory of an eclipse will be. But how do we figure out at what time the eclipse will actually reach a given point on Earth? Well, first we have to be clear on our definition of time. And there's an immediate issue with the speed of light and special relativity. What does it mean to say that the positions of the Earth and Sun are such-and-such at such-and-such a time? Because it takes light about 8 minutes to get to the Earth from the Sun, we only get to see where the Sun was 8 minutes ago, not where it is now.

And what we need is really a classic special relativity setup. We essentially imagine that the solar system is filled with a grid of clocks that have been synchronized by light pulses. And what a modern ephemeris does is to quote the results for positions of bodies in the solar system relative to the times on those clocks. (General relativity implies that in different gravitational fields the clocks will run at different rates, but for our purposes this is a tiny effect. But what isn't a tiny effect is including retardation in the equations for the n-body problem—making them become delay differential equations.)

But now there's another issue. If one's observing the eclipse, one's going to be using some timepiece (phone?) to figure out what time it is. And if it's working properly that timepiece should show official "civil time" that's based on UTC— which is what NTP internet time is synchronized to. But the issue is that UTC has a complicated relationship to the time used in the astronomical ephemeris.

The starting point is what's called UT1: a definition of time in which one day is the average time it takes the Earth to rotate once relative to the Sun. But the point is that this average time isn't constant, because the rotation of the Earth is gradually slowing down, primarily as a result of interactions with the Moon. But meanwhile, UTC is defined by an atomic clock whose timekeeping is independent of any issues about the rotation of the Earth.

There's a convention for keeping UT1 aligned with UTC: if UT1 is going to get more than 0.9 seconds away from UTC, then a leap second is added to UTC. One might think this would be a tiny effect, but actually, since 1972, a total of 27 leap seconds have been added (as specified in the Wolfram Language by GeoOrientationData):

Exactly when a new leap second will be needed is unpredictable; it depends on things like what earthquakes have occurred. But we need to account for leap seconds if we're going to get the time of the eclipse correct to the second relative to UTC or internet time.

There are a few other effects that are also important in the precise observed timing of the eclipse. The most obvious is geo elevation. In doing astronomical computations, the Earth is assumed to be an ellipsoid. (There are many different definitions, corresponding to different geodetic "datums"—and that's an issue in defining things like "sea level", but it's not relevant here.) But if you're at a different height above the ellipsoid, the cone of shadow from the eclipse will reach you at a different time. And the size of this effect can be as much as 0.3 seconds for every 1000 feet of height.

All of the effects we've talked about we're readily able to account for. But there is one remaining effect that's a bit more difficult. Right at the beginning or end of totality one typically sees points of light on the rim of the Moon. Known as Baily's beads, these are the result of rays of light that make it to us between mountains on the Moon. Figuring out exactly when all these rays are extinguished requires taking geo elevation data for the Moon, and effectively doing full 3D ray tracing. The effect can last as long as a second, and can cause the precise edge of totality to move by as much as a mile. (One can also imagine effects having to do with the corona of the Sun, which is constantly changing.)

But in the end, even though the shadow of the Moon on the Earth moves at more than 1000 mph, modern science successfully makes it possible to compute when the shadow will reach a particular point on Earth to an accuracy of perhaps a second. And that's what our precisioneclipse.com website is set up to do.

Eclipse Experiences

Written August 15, 2017

I saw my first partial solar eclipse more than 50 years ago. And I've seen one total solar eclipse before in my life—in 1991. It was the longest eclipse (6 minutes 53 seconds) that'll happen for more than a century.

There was a certain irony to my experience, though, especially in view of our efforts now to predict the exact arrival time of next week's eclipse. I'd chartered a plane and flown to a small airport in Mexico (yes, that's me on the left with the silly hat)—and my friends and I had walked to a beautiful deserted beach, and were waiting under a cloudless sky for the total eclipse to begin.

I felt proud of how prepared I was—with maps marking to the minute when the eclipse should arrive. But then I realized: there we were, out on a beach with no obvious signs of modern civilization—and nobody had brought any properly set timekeeping device (and in those days my cellphone was just a phone, and didn't even have signal there).

And so it was that I missed seeing a demonstration of an impressive achievement of science. And instead I got to experience the eclipse pretty much the way people throughout history have experienced eclipses—even if I did know that the Moon would continue gradually eating into the Sun and eventually cover it, and that it wouldn't make the world end.

There's always something sobering about astronomical events, and about realizing just how tiny human scales are compared to them. Billions of eclipses have happened over the course of the Earth's history. Recorded history has covered only a few thousand of them. On average, there's an eclipse at any given place on Earth roughly every 400 years; in Jackson, WY, where I'm planning to see next week's eclipse, it turns out the next total eclipse will be 727 years from now—in 2744.

In earlier times, civilizations built giant monuments to celebrate the motions of the Sun and Moon. Today for the eclipse next week what we're making is a website—precisioneclipse.com. But that website builds on one of the great epics of human intellectual history—stretching back to the earliest times of systematic science, and encompassing contributions from a remarkable cross-section of the most celebrated scientists and mathematicians from past centuries.

It'll be about 9538 days since the eclipse I saw in 1991. The Moon will have traveled some 500 million miles around the Earth, and the Earth some 15 billion miles around the Sun. But now—in a remarkable triumph of science—we're computing to the second when they'll be lined up again.

Written in Anticipation of April 8, 2024

In the days leading up to August 21, 2017, millions of people accessed our precisioneclipse.com website—with their geoIPs increasingly concentrating near the path of totality. I had traveled to Wyoming and—with a couple of hours to spare—found a place with a clear view across a valley. And this now being 2017, I tweeted:

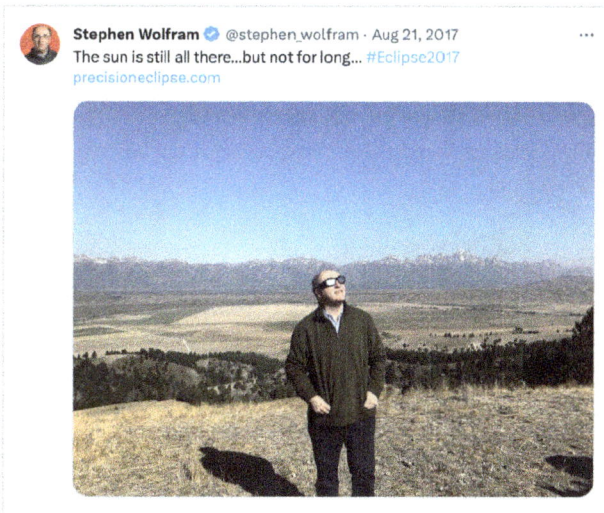

Stephen Wolfram ✔ @stephen_wolfram · Aug 21, 2017
The sun is still all there...but not for long... #Eclipse2017
precisioneclipse.com

But would our carefully done computations actually be accurate? The eclipse was going to make landfall on the Oregon coast, and, conveniently, we had a spotter right there. From where I was, the partial eclipse was well underway. And then I got a text: yes, totality in Oregon had come at the predicted second. It would take the shadow of the Moon a little under 20 minutes to reach me.

Unlike in 1991, I and everyone else had a cellphone with a precise clock—and was able to access our precisioneclipse.com site:

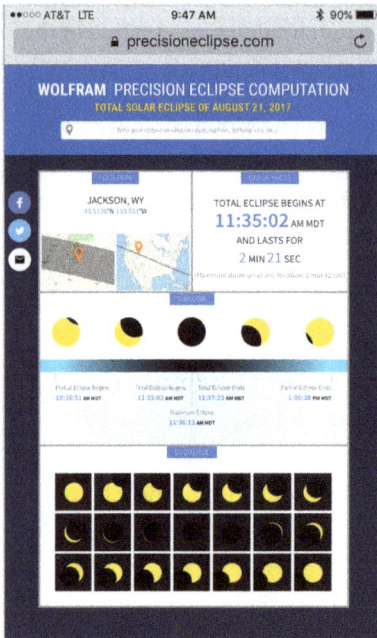

While I was waiting I was making little images of the crescent Sun with my fingers—repeating something I'd first noticed more than 50 years earlier when I saw my first partial eclipse at the age of 6:

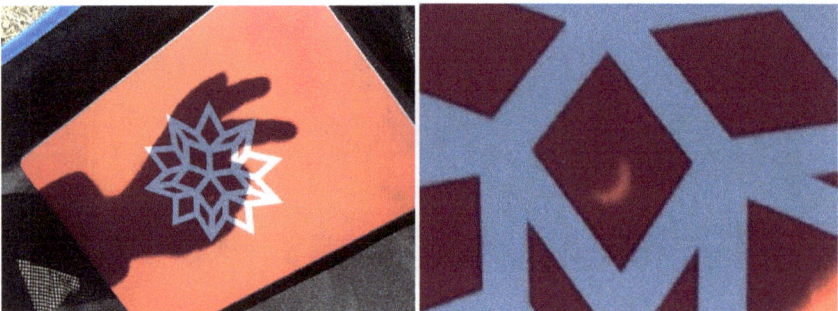

A few minutes before the eclipse, I started to see a strange shimmering (invisible on any video I took): shadow bands, a strange and poorly understood eclipse phenomenon. And then, there it came, sweeping across the valley: totality. Arriving right at the predicted second:

I had set up a camera to capture a video of the eclipse, and a little later that day I did an analysis of it—and, since earlier in 2017 I'd started routinely doing livestreams, I livestreamed it:

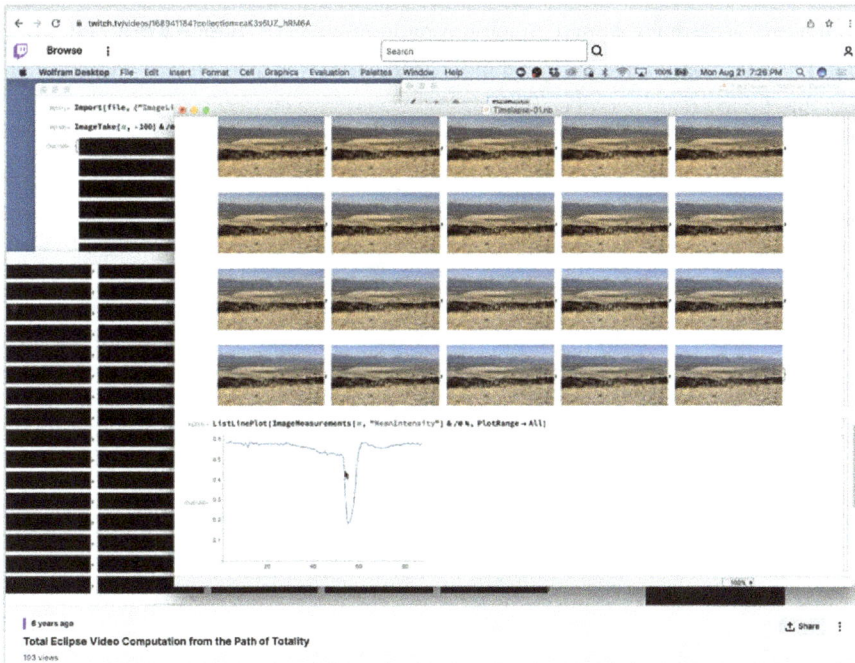

At the end, I published my notebook to the Wolfram Cloud, and it's still there:

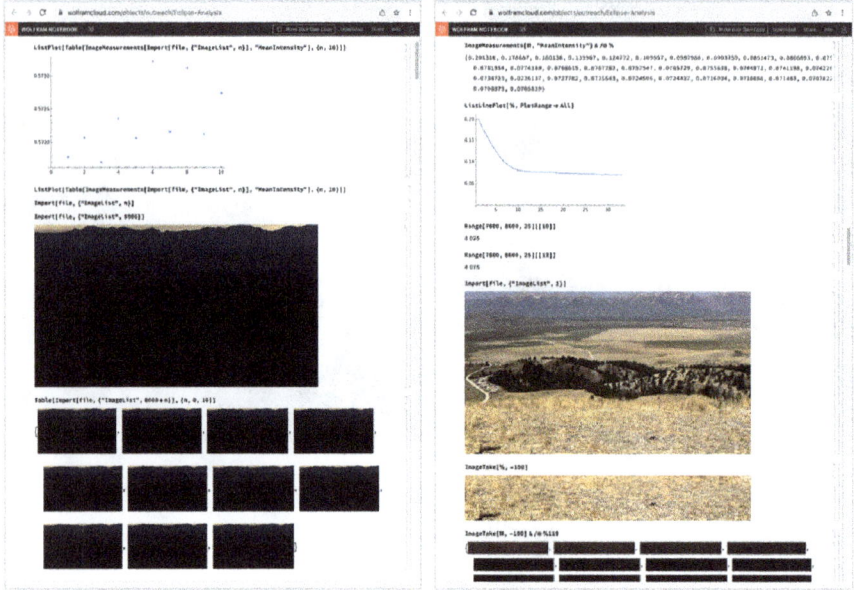

And now six years have passed. Much has happened in our human world. But the Moon has just inexorably continued in its orbit. And 2422 days later, it will once again line up to create a total eclipse...

Eclipse Computation in the Wolfram Language

The Basics

It's taken millennia to get to the point where it's possible to accurately compute eclipses. But now—as a tiny part of making "everything in the world" computable—computation about eclipses is just a built-in feature of the Wolfram Language.

The core function is SolarEclipse. By default, SolarEclipse tells us the time of the next solar eclipse from now:

In[·]:= **SolarEclipse [ ]**

Out[·]= | Mon 8 Apr 2024 11:17:20 GMT−7 |

It can also tell us the next solar eclipse from any time more than 10,000 years in the past and future:

In[·]:= **SolarEclipse [** Year: 5000 **]**

Out[·]= | Fri 9 May 5000 05:09:36 GMT−7 |

By default, SolarEclipse tells us about all solar eclipses, including partial ones. But we can request only total eclipses (or annular ones, etc.):

In[·]:= **SolarEclipse [** Year: 5000 **, EclipseType → "Total"]**

Out[·]= | Fri 9 May 5000 05:09:36 GMT−7 |

SolarEclipse immediately lets you compute nearly 100 properties of an eclipse. The most basic property is the type of eclipse; this tells us that the next eclipse from now will be total:

In[ ]:= **SolarEclipse [Now, "Type"]**

Out[ ]= total

And here are the types of all eclipses for the rest of the decade:

In[ ]:= **SolarEclipse [{** Year: 2024 **,** Year: 2030 **, All }, "Type"]**

Out[ ]= { total , annular , partial , partial , annular , total ,

annular , total , annular , total , partial , partial , partial , partial }

This gives a timeline of these eclipses; notice that most eclipses are separated by 5 or 6 months, but there's one pair (in 2029) that is just a month apart:

In[ ]:= **TimelinePlot [SolarEclipse [{** Year: 2024 **,** Year: 2030 **, All }]]**

Out[ ]=

| 2024 | 2026 | 2028 | 2030 |

Another basic property of an eclipse is its magnitude: how much of the diameter of the disk of the Sun is covered by the Moon. The next eclipse is total, so the magnitude is greater than 1:

In[ ]:= **SolarEclipse [Now, "Magnitude"]**

Out[ ]= 1.05748

Looking at the sequence of eclipses for the remainder of the decade, we see which ones are total and which are not:

In[ ]:= **SolarEclipse [{** Year: 2024 **,** Year: 2030 **, All }, "Magnitude"]**

Out[ ]= { 1.05748, 0.933371, 0.937966, 0.85533, 0.963795, 1.03953,
0.92887, 1.07998, 0.92155, 1.05694, 0.871714, 0.457571, 0.230097, 0.891385 }

Here's the path of totality for the upcoming total eclipse:

In[ ]:= **GeoListPlot [SolarEclipse [Now, "UmbraEnvelopePolygon"]]**

Out[ ]=

And here's where the eclipse is partial:

In[ ]:= **GeoListPlot [SolarEclipse [Now, "PenumbraEnvelopePolygon"]]**

Out[ ]=

This is easier to understand with a different geo projection:

In[ ]:= **GeoListPlot [SolarEclipse [Now, "PenumbraEnvelopePolygon"] ,
GeoProjection → "Orthographic"]**

Out[ ]=

There are all sorts of special points and lines associated with these regions. This gives the "point of maximum eclipse" (i.e. essentially the place where the eclipse lasts longest):

```
In[ ]:= SolarEclipse [ Now, "MaximumEclipsePosition" ]
```

```
Out[ ]= GeoPosition [ { 25.2896, −104.148 } ]
```

And this gives the precise time (converted to my current time zone) of the maximum eclipse:

```
In[ ]:= SolarEclipse [ Now, "MaximumEclipseDate" ]
```

Out[ ]= Mon 8 Apr 2024 11:17:20 GMT−7

This gives the time of maximum eclipse in the time zone of the point of maximum eclipse:

```
In[ ]:= LocalTime [ SolarEclipse [ Now, "MaximumEclipsePosition" ] ,
          SolarEclipse [ Now, "MaximumEclipseDate" ] ]
```

Out[ ]= Mon 8 Apr 2024 12:17:20 GMT−6

At the point of maximum eclipse, this gives the duration of the umbra (i.e. the time of totality):

```
In[ ]:= SolarEclipse [ Now, "MaximumEclipseUmbraDuration" ]
```

```
Out[ ]= 4.53535 min
```

And here's a map of where on the Earth the umbra is at the time of maximum eclipse:

```
In[ ]:= GeoListPlot [ SolarEclipse [ Now,
          {"UmbraPolygon", SolarEclipse [ Now, "MaximumEclipseDate" ] } ] ]
```

Out[ ]=

Zooming out to a range of 500 miles makes it easier to tell where this is:

In[ ]:= **GeoListPlot [SolarEclipse [Now, { "UmbraPolygon",**
       SolarEclipse [Now, "MaximumEclipseDate"] }] , GeoRange → 500 mi]

Out[ ]=

This shows the position of the umbra every minute for the hour after the time of maximum eclipse:

In[ ]:= **GeoListPlot [Table [SolarEclipse [Now,**
       { "UmbraPolygon", SolarEclipse [Now, "MaximumEclipseDate"] +
         Quantity [m, "Minute"] }] , { m, 60 }]]

Out[ ]=

Here's a summary map of the eclipse, including for example (in red) the points of
first and last contact of the penumbra:

In[◦]:= **SolarEclipse [Now, "EclipseMap"]**

Out[◦]=

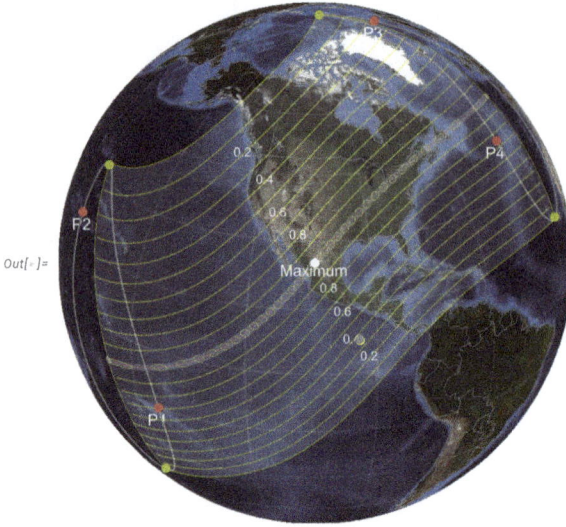

Instead of looking at what amounts to the shadow of the Moon on the Earth, we
can ask what we'd see in the sky. During a total eclipse the Moon will completely
cover the Sun. But here's what happens just 15 minutes before the time of maximum eclipse:

In[◦]:= **AstroGraphics [** Sun STAR ,

   AstroReferenceFrame → { SolarEclipse [Now, "MaximumEclipseDate"] − 15 min ,

   SolarEclipse [Now, "MaximumEclipsePosition"] }, AstroRange → 1°]

Out[◦]=

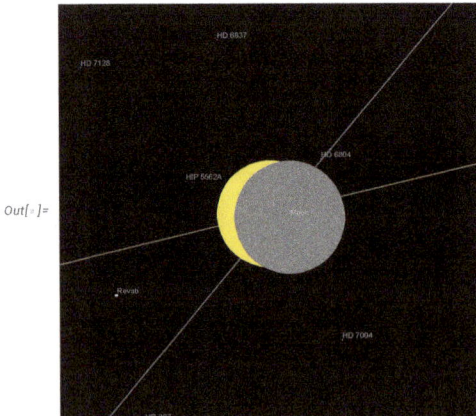

Lots of Eclipses

Here's a map of the path of totality for all the total eclipses for the next 50 years:

```
In[ ]:= GeoListPlot [ SolarEclipse [ { Now, Now + Quantity [ 50, "Years" ] , All } ,
            "UmbraEnvelopePolygon", EclipseType → "Total" ] ,
        PlotStyle → Red, GeoProjection → "Robinson" ]
```

Out[ ]=

Over the course of 500 years there are lots of total eclipses:

```
In[ ]:= GeoListPlot [ SolarEclipse [ { Now, Now + Quantity [ 500, "Years" ] , All } ,
            "UmbraEnvelopePolygon", EclipseType → "Total" ] ,
        PlotStyle → Red, GeoProjection → "Robinson" ]
```

Out[ ]=

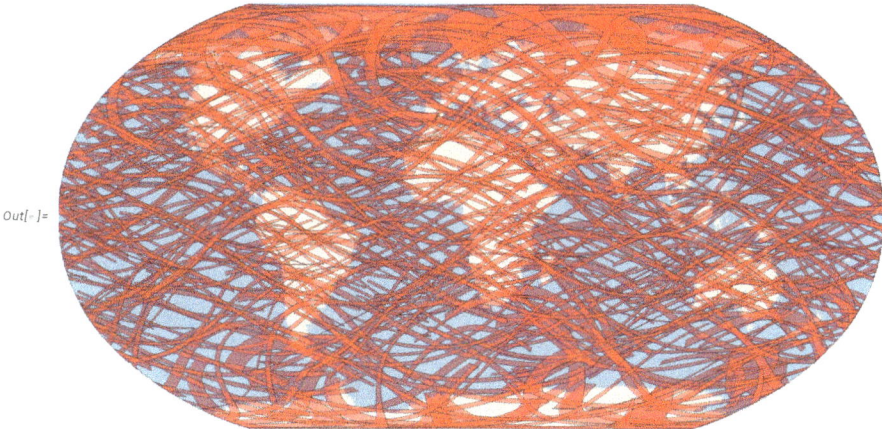

Although it's not terribly obvious there, there's actually a lot of regularity in these paths. In particular, as we discussed previously, similar eclipses occur in "saros series", separated by a time of about 1 saros, or roughly 18 years. Here are the paths of eclipses that appear for the 10 saroses after the next eclipse:

In[ ]:= **GeoListPlot [Table [SolarEclipse [SolarEclipse [ ] + Quantity [s, "Saroses"] ,**
   "UmbraEnvelopePolygon", EclipseType → "Total"] , { s, 10 }] ,
   PlotStyle → Red, GeoProjection → "Robinson"]

Out[ ]=

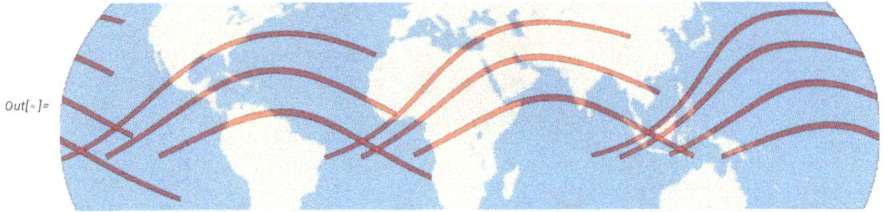

Each successive eclipse in the saros series is systematically about 120° to the west of the previous one, and a little south (or north, depending on the series). The series continues like this until the eclipse paths hit one of the poles, at which point the series ends:

In[ ]:= **GeoListPlot [Table [SolarEclipse [SolarEclipse [ ] + Quantity [s, "Saroses"] ,**
   "UmbraEnvelopePolygon", EclipseType → "Total"] , { s, 25 }] ,
   PlotStyle → Red, GeoProjection → "Robinson"]

Out[ ]=

Any given eclipse is in a saros series. The next eclipse is in series 139 (the numbering scheme for these series was set in 1955—with series 0 chosen, quite arbitrarily, to be the one that starts just after 3000 BC):

In[ ]:= **SolarEclipse [Now, "SarosSeries"]**

Out[ ]= 139

There are 71 eclipses in this saros series

In[ ]:= **Length [SolarEclipse [{ "Saros", 139 }]]**

Out[ ]= 71

running from the year 1501 to the year 2763 (a span of 1263 years):

In[ ]:= **TimelinePlot [SolarEclipse [{ "Saros", 139 }]]**

Out[ ]=

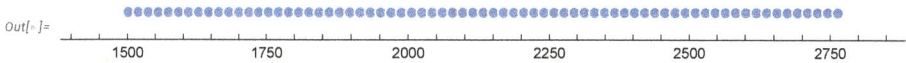

Not all these eclipses are total, however. But if we plot the magnitudes of all the eclipses, we see that the partial ones appear only at the ends of the saros series:

In[ ]:= **ListPlot [SolarEclipse [SolarEclipse [{ "Saros", 139 }] , "Magnitude"]]**

Out[ ]=

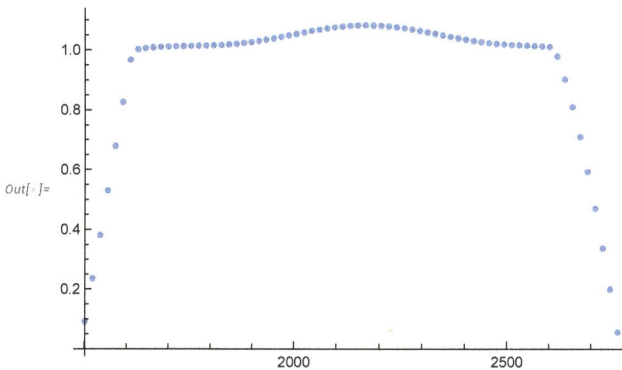

If we look at the next few eclipses, we'll see that they are part of all sorts of different saros series:

In[ ]:= **SolarEclipse [{ Year: 2024 , Year: 2030 , All}, "SarosSeries"]**

Out[ ]= { 139, 144, 149, 154, 121, 126, 131, 136, 141, 146, 151, 118, 156, 123 }

Here are dates for eclipses in a sequence of different saros series:

```
In[ ]:= DateListPlot [
           Table [ Thread [ { SolarEclipse [ { "Saros", s } ] , s } ] , { s, 110, 140 } ] , Joined → False ]
```

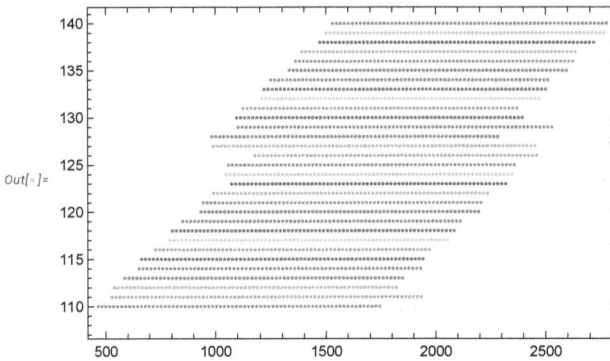

Right now there are 40 saros series active:

```
In[ ]:= Select [ Range [ 110, 160 ] , s ↦
           DateWithinQ [ DateInterval [ DateBounds [ SolarEclipse [ { "Saros", s } ] ] ] , Now ] ]
```

```
Out[ ]= {117, 118, 119, 120, 121, 122, 123, 124, 125, 126, 127, 128, 129, 130, 131, 132, 133, 134, 135, 136,
          137, 138, 139, 140, 141, 142, 143, 144, 145, 146, 147, 148, 149, 150, 151, 152, 153, 154, 155, 156}
```

Any given eclipse could be specified by its "index number" within its saros series. But in the mid-twentieth century it was realized that there's a more convenient and robust way to label eclipses, using a combination of "saros number" and what's called "inex number".

As we discussed previously, for an eclipse to occur, a new moon must happen at a time when the Moon is close to the plane of the ecliptic. The average time between new moons is the so-called synodic month:

```
In[ ]:= synodic = UnitConvert [ 1 synodic month , "Days" ]
```

```
Out[ ]= 29.5305889050 days
```

Meanwhile, the average time between the Moon's crossings of the plane of the ecliptic is half the so-called draconic month:

```
In[ ]:= draconic = UnitConvert [ 1 Draconian month , "Days" ]
```

```
Out[ ]= 27.2122199 days
```

Given a particular eclipse, the time before an approximate "repeat eclipse" will correspond to a coincidence between an integer multiple of the synodic month, and of half the draconic month. And to figure out when these coincidences will occur is essentially a question of number theory.

Let's compute the continued fraction expansion:

```
In[ ]:= cf = ContinuedFraction [ synodic / (draconic / 2) ]
Out[ ]= {2, 5, 1, 6, 1, 1, 1, 1, 1, 10, 1, 1}
```

From this we can derive a sequence of rational approximations:

```
In[ ]:= rat = Table [ FromContinuedFraction [ Take [ cf, n ] ], {n, 10} ]
```

$$Out[ ]= \left\{2, \frac{11}{5}, \frac{13}{6}, \frac{89}{41}, \frac{102}{47}, \frac{191}{88}, \frac{293}{135}, \frac{484}{223}, \frac{777}{358}, \frac{8254}{3803}\right\}$$

These approximations get progressively better:

```
In[ ]:= rat - synodic / (draconic / 2)
```

$$Out[ ]= \{-0.17039176, 0.02960824, -0.00372509, 0.00033995, -0.00017899, 0.00006279, -0.00002139, 0.00001183, -7.0 \times 10^{-7}, 4. \times 10^{-8}\}$$

The 8th one is 484/223—and this corresponds to the saros cycle, which reflects the close similarity of 223 synodic months and 242 draconic months:

```
In[ ]:= {484 draconic / 2, 223 synodic}
Out[ ]= { 6585.35722 days , 6585.32132581 days }
```

But now let's look at the 9th rational approximation: 777/358. This reflects the coincidence:

```
In[ ]:= {777 draconic / 2, 358 synodic}
Out[ ]= { 10571.9474 days , 10571.9508280 days }
```

And now this coincidence defines another cycle—which is the inex cycle. There are lots of other cycles one can identify—but all the common ones can be expressed as linear combinations of the saros and inex cycles.

We saw above how eclipses occur in saros series. But we now see that they will also occur in inex series. And a convenient way to specify an eclipse is to say what saros series and what inex series it appears in. The saros and inex series are numbered according to when they started, with the 0th saros series by convention being the one that spans:

In[ ]:= **DateBounds [SolarEclipse [{ "Saros", 0 }]]**

Out[ ]= { Sun 27 Apr –2956 23:00:13 GMT–7 , Sun 14 Jun –1676 10:04:47 GMT–7 }

With this setup, the April 2024 eclipse can then be specified by its saros and inex numbers:

In[ ]:= **SolarEclipse [Now, "SarosInex"]**

Out[ ]= { 139, 55 }

But how does this fit in with other eclipses? Here's a plot of the inex and saros numbers of all eclipses between 1000 AD and 3000 AD (with the April 2024 eclipse indicated in red):

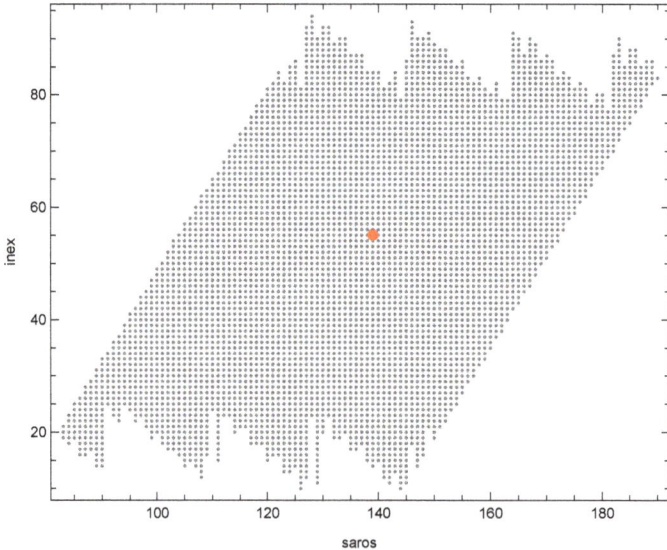

Each saros series shows up as a vertical line of eclipses, and each inex series as a horizontal line. The finite overall date range for the picture leads to the diagonal cutoffs on each side.

For each saros, inex number pair there'll be some kind of eclipse. But most of the eclipses won't be total. Here's where total eclipses occur in the saros, inex plane:

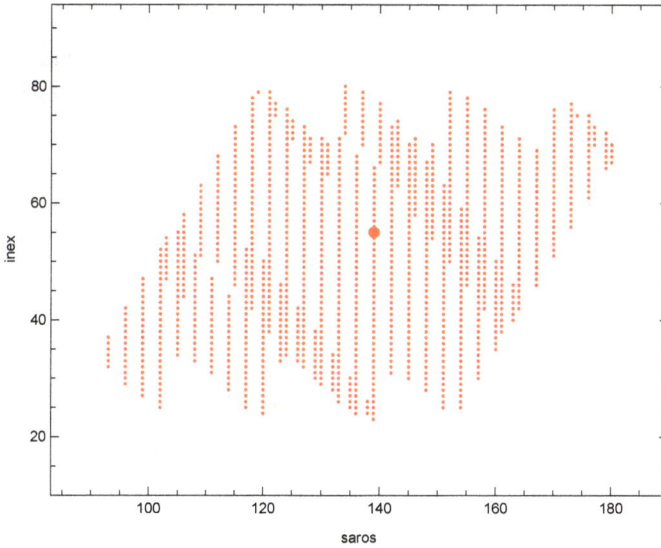

And here's a complete color-coded "panorama" of different kinds of eclipses:

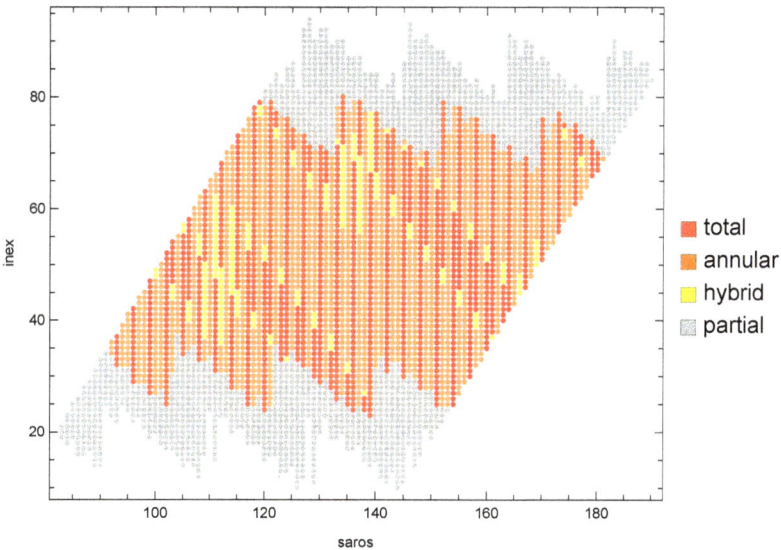

(Note that—as we discussed above—the eclipses at the ends of saros series are partial.)

The Eclipse from First Principles

The function SolarEclipse in the Wolfram Language immediately tells us when eclipses occur. But we can also deduce this information from other "lower-level" functions. In particular, we know that an eclipse (of at least some kind) occurs if the angular separation between the Sun and the Moon in the sky is smaller than the total of their apparent radii (or about 0.5°). The angular separation depends on where you are on the Earth. Let's pick a location from which we know an eclipse will be visible:

In[ ]:= **loc = SolarEclipse ["MaximumEclipsePosition"]**

Out[ ]= GeoPosition [{ 25.2896, –104.148 }]

Now let's plot angular separation for each hour over the course of the next year:

In[ ]:= **DateListPlot [Table [**

    { t, AstroAngularSeparation [Sun STAR **, Moon** PLANETARY MOON **, { t, loc }] },**

    { t, Year: 2024 ["BeginningInstant"] , Year: 2024 ["EndInstant"] , 1 h }]]

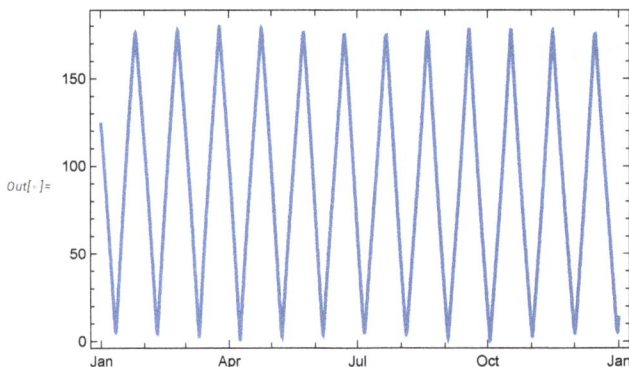

The minima occur once per lunar month, near the time of the new moons. Changing the plot range, we can see that these minima are different in different (lunar) months:

In[ ]:= **Show [%, PlotRange → { 0, 20 }]**

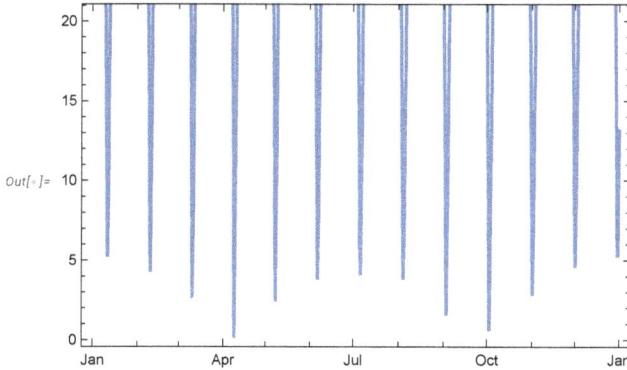

Let's look at early April in more detail, now sampling every minute:

In[ ]:= **DateListPlot [Table [**

    { t, AstroAngularSeparation [Sun STAR **,** Moon PLANETARY MOON **, { t, loc }] },**

    { t, Fri 5 Apr 2024 00:00:00 **,** Fri 12 Apr 2024 00:00:00 **, 1 min }]]**

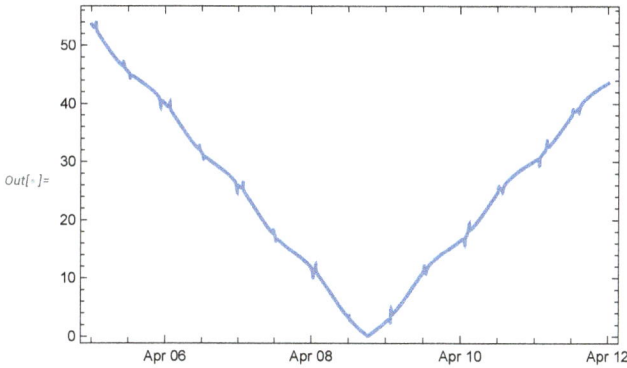

And what we see here is that the angular separation goes to zero—reflecting the fact that there's a total eclipse. What are those little glitches? They're a consequence of the fact that the apparent positions of the Sun and Moon change when they're close to the horizon because of refraction in the atmosphere. Not including refraction gives a smoother curve:

In[ ]:= **DateListPlot [Table [{ t, AstroAngularSeparation [`Sun` STAR ,**

`Moon` PLANETARY MOON , { t, loc, "Refraction" → None }] },

{ t, `Fri 5 Apr 2024 00:00:00` , `Fri 12 Apr 2024 00:00:00` , 1 min }]]

Out[ ]=

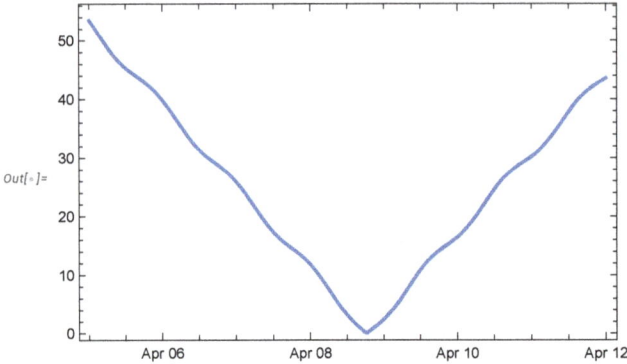

We've been looking at the angular separation between the centers of the disks of the Sun and Moon in the sky. But what about the actual positions in the sky? Here's the astro position of the Sun (in local horizon coordinates) at the time and place of maximum eclipse:

In[ ]:= **AstroPosition [`Sun` STAR ,**

{ SolarEclipse [ ] , SolarEclipse ["MaximumEclipsePosition"] }]

Out[ ]= AstroPosition[
Azimuth: 149° 23' 13.8"
Altitude: 69° 47' 50.1"
]

And here now is the almost-exactly-equal result for the Moon:

In[ ]:= **AstroPosition [`Moon` PLANETARY MOON ,**

{ SolarEclipse [ ] , SolarEclipse ["MaximumEclipsePosition"] }]

Out[ ]= AstroPosition[
Azimuth: 149° 23' 13.7"
Altitude: 69° 47' 50.1"
]

What will it actually look like in the sky? Here's now a graphic showing the disks of the Sun and Moon 15 minutes before the time of maximum eclipse:

In[◦]:= **AstroGraphics [** Sun STAR **, AstroReferenceFrame → { SolarEclipse [ ] − 15 min ,**

SolarEclipse ["MaximumEclipsePosition"] }, AstroRange → 0.5°]

Out[◦]=

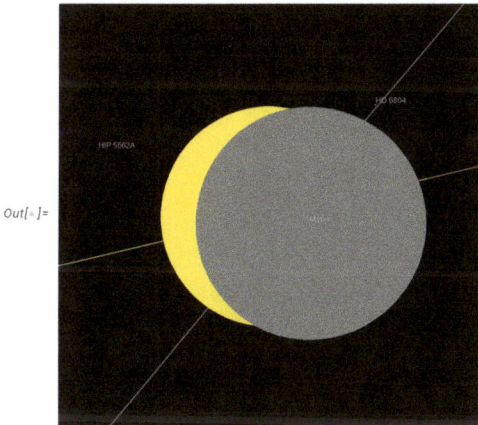

Here's a sequence of images 15 minutes apart:

In[◦]:= **Table [AstroGraphics [** Sun STAR **, AstroReferenceFrame →**

{ SolarEclipse [ ] + n 15 min , SolarEclipse ["MaximumEclipsePosition"] },

AstroRange → 0.5° , ImageSize → Tiny], { n, −5, 5 }]

Out[◦]=

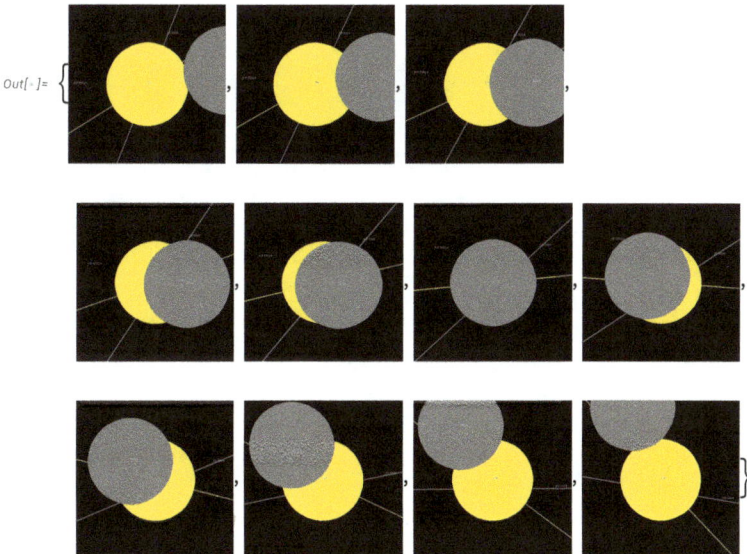

Analyzing Eclipse Data

The 2017 Eclipse

I saw the 2017 eclipse from a rather scenic spot near Jackson, Wyoming:

Specifically (according to Wolfram|Alpha accessed through my phone) I was at geo position 43.5125°N 110.6506°W—at an elevation of 7526 ft:

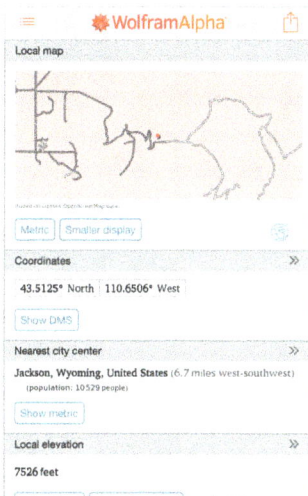

As a check, here's the expected (ground) elevation at that lat-lon—definitely within bounds, particularly considering I was holding the phone about 5 feet off the ground, etc.:

In[ ]:= **GeoElevationData [GeoPosition [{ 43.5125, −110.6506 }]]**

Out[ ]= 7513.12 ft

Defining my location as

In[ ]:= **location = GeoPosition [{ 43.5125, −110.6506, 7526 ft }];**

the predicted arrival time of the eclipse at that location was then:

In[ ]:= **time = SolarEclipse [Aug 2017 , { "LocalUmbraContact1Date", location },**

      TimeZone → LocalTimeZone [Jackson CITY]]

Out[ ]= Mon 21 Aug 2017 11:35:01 MDT

And indeed that's exactly what our precisioneclipse.com site told me at the time:

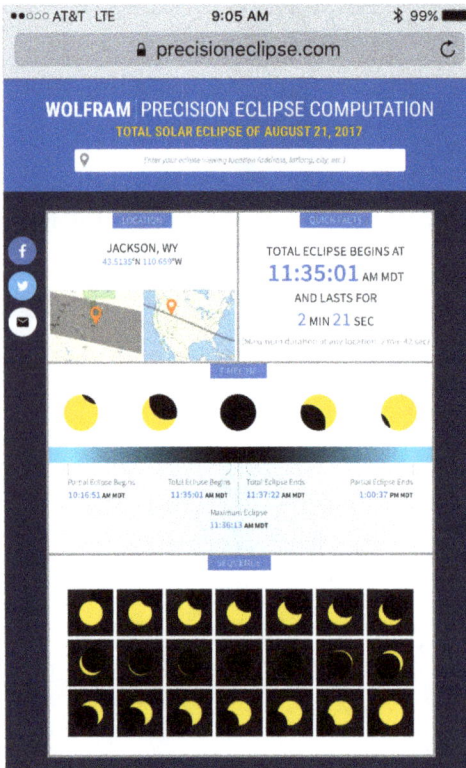

What would happen at the "moment of totality"? Here's where the shadow of the Moon was predicted to be (where I was standing is indicated by a red dot; the whole picture is 100 miles across):

In[ ]:= **GeoGraphics [{ SolarEclipse [** Aug 2017 **, { "UmbraPolygon", time }] ,**
        Style [Point [location] , PointSize [.02] , Red] } ,
        GeoBackground → "ReliefMap", GeoRange → 50 mi **]**

Out[ ]=

Fifteen seconds earlier, the shadow of the Moon would have just "crossed" the row of mountains I could see:

In[ ]:= **GeoGraphics [{ SolarEclipse [** Aug 2017 **, { "UmbraPolygon", time −** 15 s **}],**

Style [Point [location] , PointSize [0.02] , Red] }, ⋯ ✛ **]**

Out[ ]=

Looking (in exaggerated 3D) at the terrain, here's the "umbral cone" at the moment of totality for me:

And it was moving at a speed

In[•]:= **UnitConvert [SolarEclipse [** Aug 2017 **,**
{ **"ShadowAxisSpeed",** Mon 21 Aug 2017 11:35:03 MDT **}] ,** mi / h **]**

Out[•]:= 1860.17 mi / h

which is effectively a vector difference of the rotation speed of the Earth

In[•]:= **Earth** PLANET **[** equatorial velocity **]**

Out[•]= 1040.4014 mi / h

corrected for latitude

$In[\circ]:=$ **Cos [Latitude [location]]** $\boxed{\textbf{Earth} \text{ PLANET}}$ **[** $\boxed{\textit{equatorial velocity}}$ **]**

$Out[\circ]=$ 754.523 mi / h

and the orbital velocity of the Moon:

$In[\circ]:=$ $\boxed{\textbf{Moon} \text{ PLANETARY MOON}}$ **[** $\boxed{\textit{average orbit velocity}}$ **]**

$Out[\circ]=$ 2.28×10^3 mi / h

Shown at 5-second intervals for 30 seconds, here's how the edge of the umbra was moving just before it reached me:

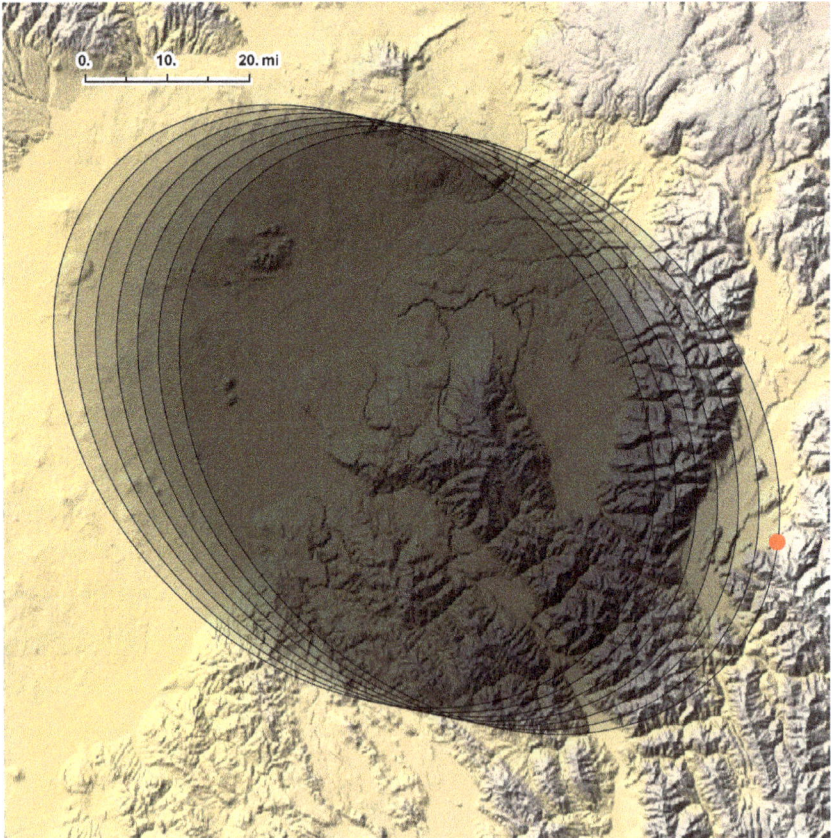

That day, I had brought some not-very-high-tech "equipment" to record the eclipse:

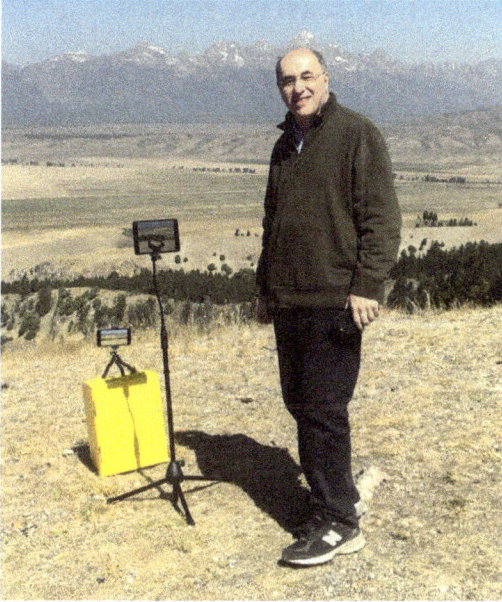

The main video I captured—which was from the iPad—is now stored for posterity in the Wolfram Data Repository:

In[·]:= **video = Video [ResourceData ["Videos of the 2017 Solar Eclipse", "LongVideo"]]**

Out[·]=

Scan to watch online

▶ 00:00   11:00 ◀)) ≡

It was 11 minutes long. Sampling frames from it we get:

In[∘]:= **ImageAssemble [Partition [VideoFrameList [video, 30] , 6]]**

Out[∘]=

To get more of a sense of the eclipse, we can pick out the center column of the video, and arrange it in time:

In[∘]:= **Show [ImageAssemble [VideoMapList [ImageCrop [#Image, {1, Full}] &, video]] , AspectRatio → 1 / 4]**

Out[∘]=

Notice that the "band of totality" appears slightly further to the left at the top of the image—reflecting the fact that totality reaches the mountains in the distance (at the top of the image) slightly earlier than it reaches the "foreground" at the bottom of the image.

To be more quantitative, we can measure the mean intensity of the bottom part of the image as a function of time (where here we've aligned with the timestamp that records the start time of the video):

```
In[ ]:=  Block [ { $TimeZone = -6 } , DateListPlot [ Parallelize @ VideoMapList [
           ImageMeasurements [ ImageTake [ #Image, -50 ] , "MeanIntensity" ] &,
           video ] , {  Mon 21 Aug 2017 11:30:33 MDT  ,  Mon 21 Aug 2017 11:41:33 MDT  } ,
           AspectRatio → 1 / 3 ] ]
```

Out[ ]:=

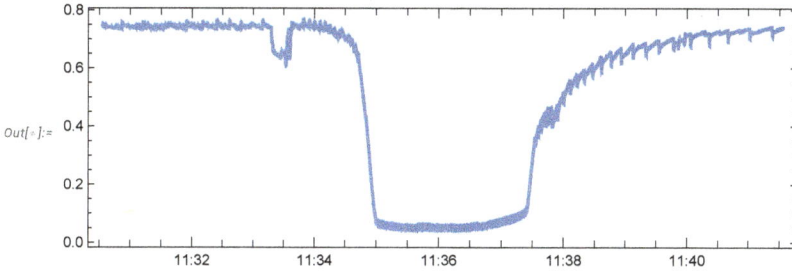

We see in this "light curve" a big dip during the period of totality. There's also a little dip earlier from someone walking in front of the camera. And we also see lots of little glitches that we'll discuss later.

But what should we expect this light curve to look like? Well, we can predict what the obscuration of the Sun around the time of totality should be—and but for the 20% effect of "limb darkening", this should give the intensity of sunlight:

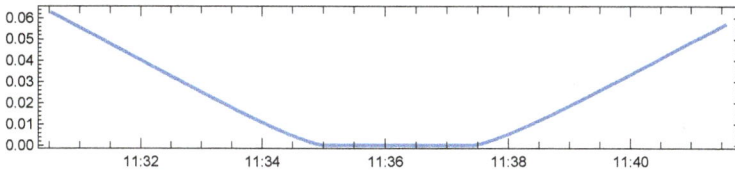

Rescaling values somewhat, we can compare the curves:

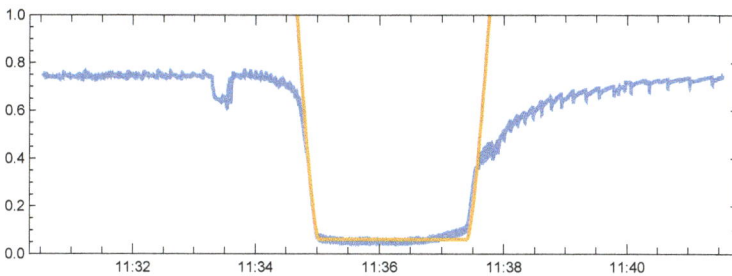

And zooming in on the minute around the beginning of totality, we see:

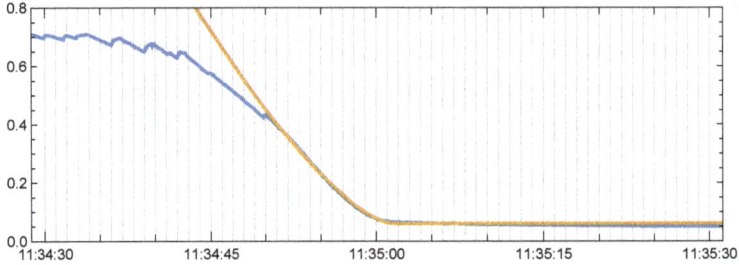

And, yes indeed, the observed onset of totality seems to agree basically to the second with our prediction!

But even though there's this agreement, the overall shapes of the observed and predicted light curves definitely aren't the same. And, yes, this is a story of experimental science. And, arguably, of a mistake I made—of using consumer electronics, optimized for consumer purposes, to make a quantitative scientific measurement. You see, as I now realize, an iPad by default always tries to "get a good picture" by maintaining the brightness of an image independent of overall light level. And while for "consumer purposes" that's usually the right thing to do, it definitely confuses things if one's trying to measure the light curve for an eclipse.

And indeed if we look at our "measured light curve" it's very flat until the period of totality. In other words, the iPad succeeded in maintaining the same image brightness until there just wasn't enough light at all, at which point the image "faded to black". (At the end of totality, the iPad gradually realized "yes, there's more light now", and the measured light curve slowly climbs back up.)

But what's with all the glitches we see? They're already visible in our "video time collage" above. And one thought might be that they're an actual eclipse phenomenon—perhaps associated with the "shadow bands" that I did indeed see just before this eclipse. But the characteristic shimmering associated with such shadow bands—while very difficult to capture on video, perhaps because actual images aren't being formed—happens much faster than the glitches we're seeing.

Looking in a bit more detail, we see that there are upward glitches in the period before totality, and downward ones after. Zooming in on a couple of minutes of the "before" period and a couple of minutes of the "after" period, we see:

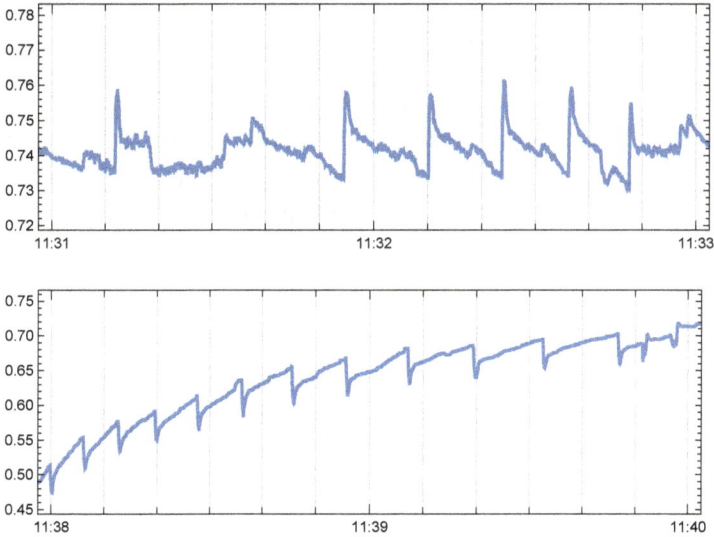

And, yes, we can validate that this is an "instrumental phenomenon" by doing a simple experiment—using the very same iPad as in 2017—and continuously sliding a piece of cardboard in front of a light and then capturing video and measuring the light curve for this in-my-basement-style "model eclipse":

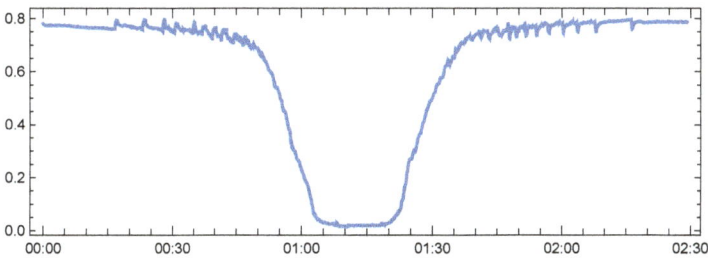

We see the very same kind of glitches as from the actual eclipse video. And presumably in both cases they're reflections of the automatic exposure control system used by the iPad. (If the individual frames of the video had EXIF metadata, we might be able to see that explicitly, but as it is, there's only EXIF data for the whole video.) I don't know in detail how this iPad's exposure control system works, but what we're seeing is a kind of overshooting-and-correction that's very common in all sorts of control systems. If we knew in advance everything that happens in the video, then maybe we could avoid the glitches. But if we're going to try to maintain light level on an ongoing basis during the recording of the video (perhaps by adjusting the gamma correction that determines how raw sensor values are translated to pixel values), then control theory most likely implies that glitches are inevitable.

My First Eclipse

So many years later, I still remember it well. I was 6 years old (almost 7), walking the couple of blocks to school (yes, on my own, which kids in England in those days did). It happened when I was walking under a tree (and, yes, I still remember exactly where). There were lots of dappled patches of light on the ground. And something looked strange about them. And suddenly I realized what it was—and I still have an image of it in my mind today. All the patches of light had the same bite taken out of them. And it didn't take me long to look up at the Sun, and see that it too had a bite taken out of it.

Being already something of a science enthusiast, I'd heard of eclipses, and realized this must be one. I arrived at school a few minutes later, and regaled the other kids with my "discovery". But despite the obviousness (at least to me) of what was going on, I wasn't widely believed. And, yes, that was a very educational experience. But that's a different story…

So what was that eclipse? Well, it was the one in May 1966:

In[ ]:= **SolarEclipse [** Year: 1966 **, TimeZone →** Europe/London TIME ZONE **]**

Out[ ]= Fri 20 May 1966 10:38:24 BST

It was a partial eclipse—visible from England:

In[ ]:= **SolarEclipse [** Year: 1966 **, "EclipseMap"]**

Out[ ]=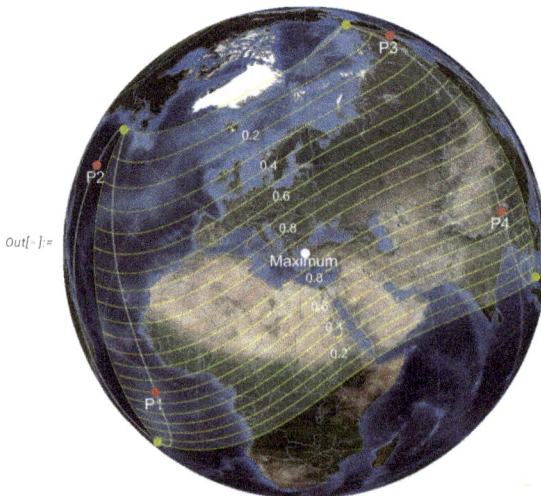

The geo location of my "discovery tree" was 51.7636° N 1.2558° W. So now we can compute the magnitude of the eclipse there that morning:

In[ ]:= **DateListPlot [Table [{t, SolarEclipse [** Year: 1966 **,**

{"LocalMagnitude", t, GeoPosition [{51.7636, −1.25583}] },

TimeZone → Europe/London TIME ZONE **] },**

{t, Fri 20 May 1966 07:00:00 **,** Fri 20 May 1966 11:00:00 **, 5 min }]]**

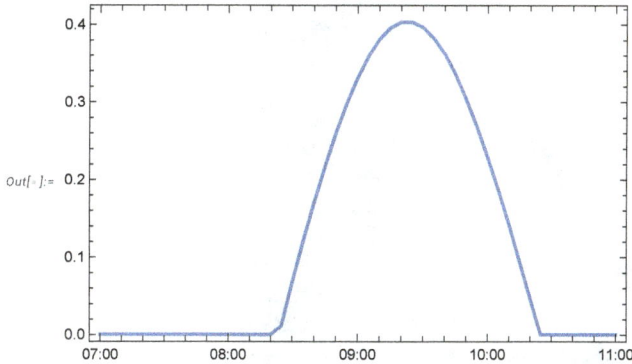

I believe school started at 9am, so what I saw was an eclipse with the rough magnitude:

In[ ]:= **SolarEclipse [** Year: 1966 **,**

{"LocalMagnitude", Fri 20 May 1966 08:55:00 **, GeoPosition [{51.7636, −1.25583}] },**

TimeZone → Europe/London TIME ZONE **]**

Out[ ]:= 0.297351

The Sun (and Moon) were about 50° above the horizon:

In[ ]:= **SunPosition [GeoPosition [{51.7636, −1.25583}] ,** Fri 20 May 1966 08:55:00 **]**

Out[ ]:= { 225.564° , 51.2624° }

In[ ]:= **MoonPosition [GeoPosition [{51.7636, −1.25583}] ,** Fri 20 May 1966 08:55:00 **]**

Out[ ]:= { 223.441° , 52.4799° }

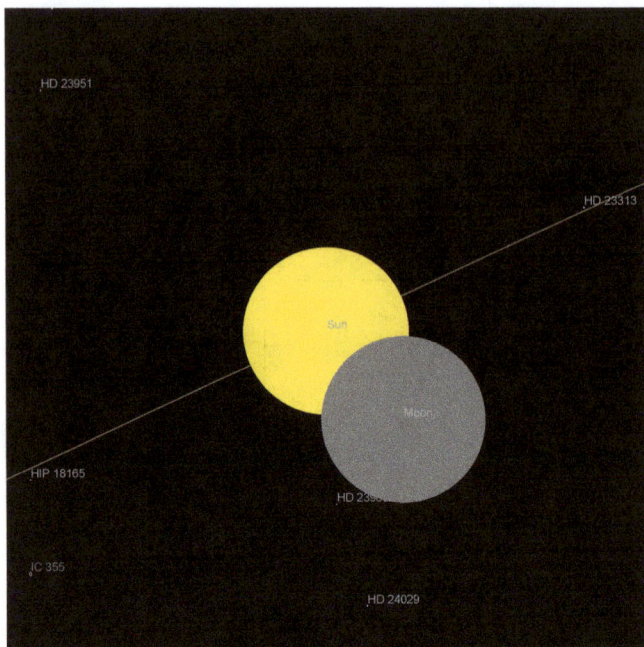

And the Moon was poking into the disk of the Sun:

In[ ]:= **ResourceFunction ["SolarEclipseIcon"] [**
　　　　　Fri 20 May 1966 08:55:00 **, GeoPosition [{ 51.7636, −1.25583 }]]**

Out[ ]:=

Thirty minutes later the Moon had poked a little further into the disk of the Sun. But after a bit more than a hour, the whole Sun was back, and the eclipse was over:

In[ ]:= **GraphicsRow [Table [ResourceFunction ["SolarEclipseIcon"] [**
　　　　　Fri 20 May 1966 08:55:00 **+ Quantity [t, "Minutes"] ,**
　　　　　GeoPosition [{ 51.7636, −1.25583 }]] , { t, 0, 90, 10 }]]

Out[ ]:=

And it would be 25 years before I'd see another eclipse—though this time a total one.

By the way, this is me back at the time of my first eclipse—captured in a long-before-those-were-popular "selfie", taken with a film camera and manual focus, and, it seems, a lot of concentration (and, yes, I think I still make the same strange expression when I'm concentrating hard today):

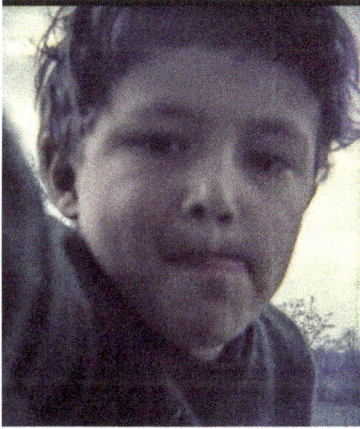

Maps for the Eclipse
of April 8, 2024

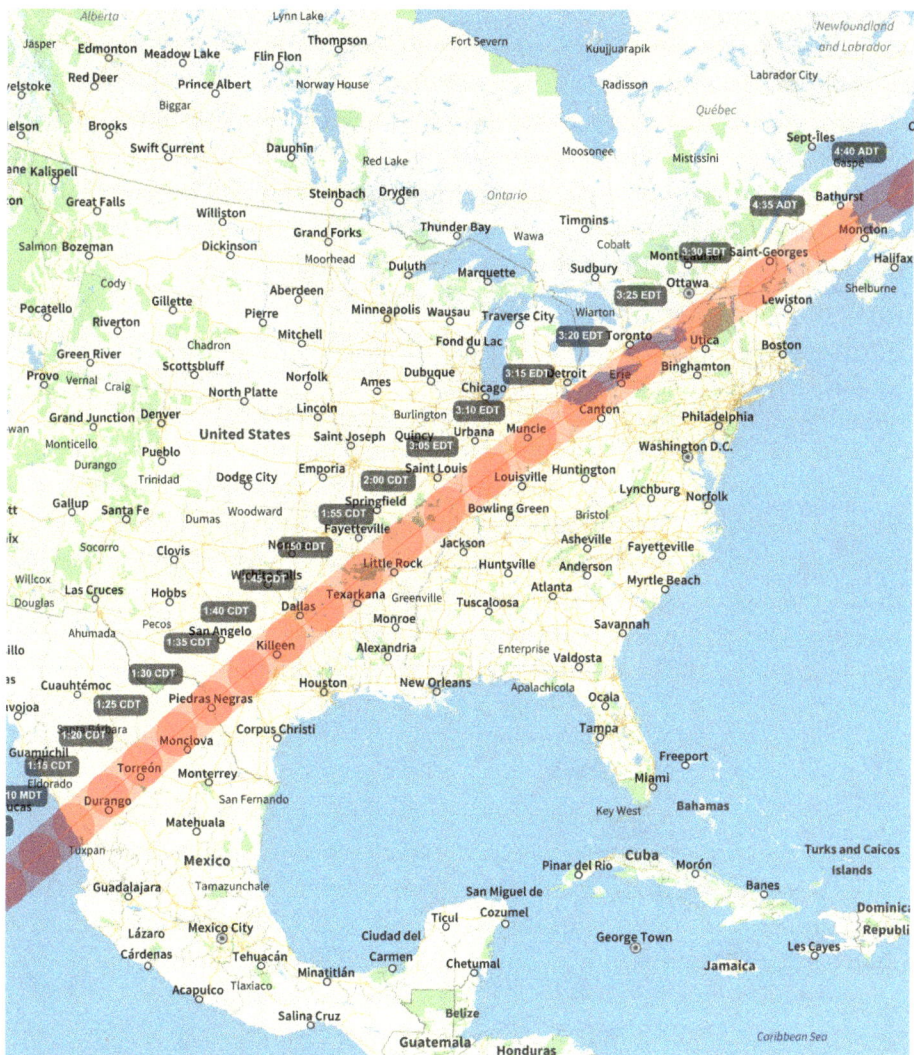

Mazatlán/Durango

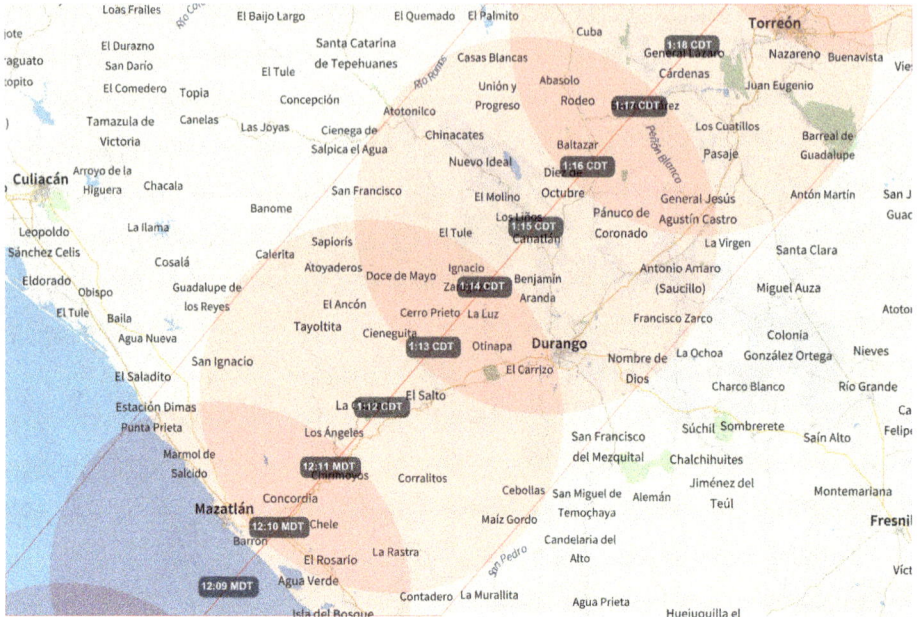

Torreón

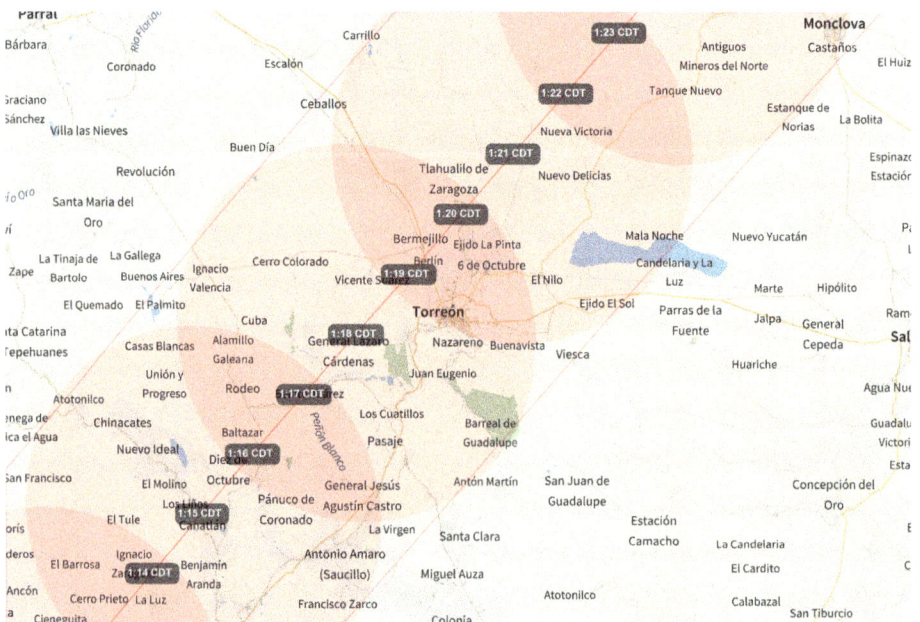

San Antonio/Austin

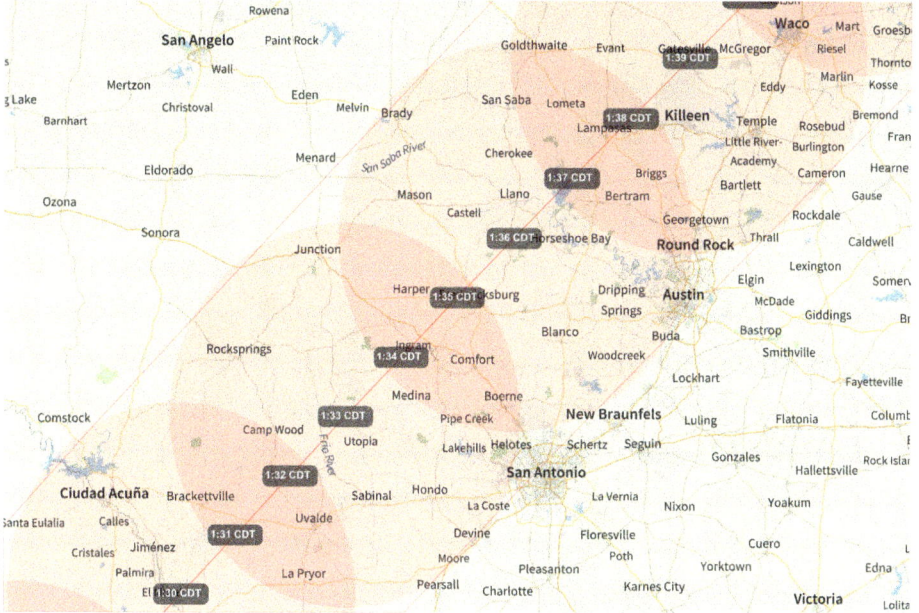

Dallas/Fort Worth

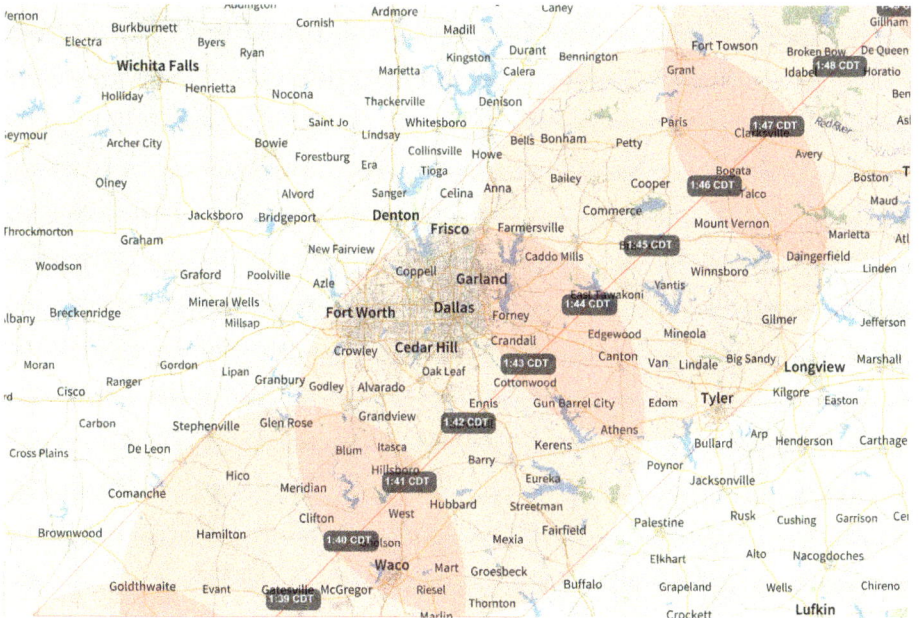

St. Louis

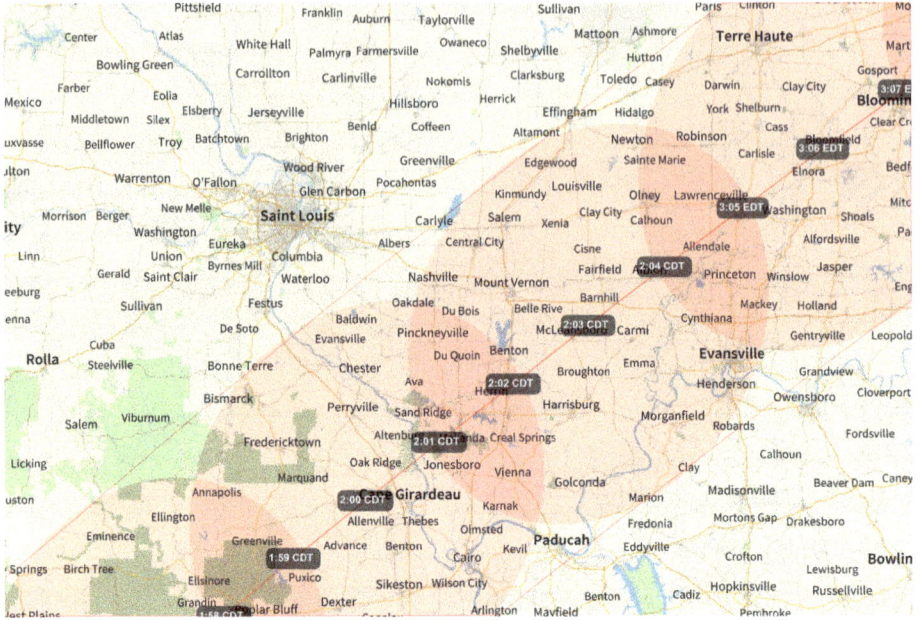

Indianapolis

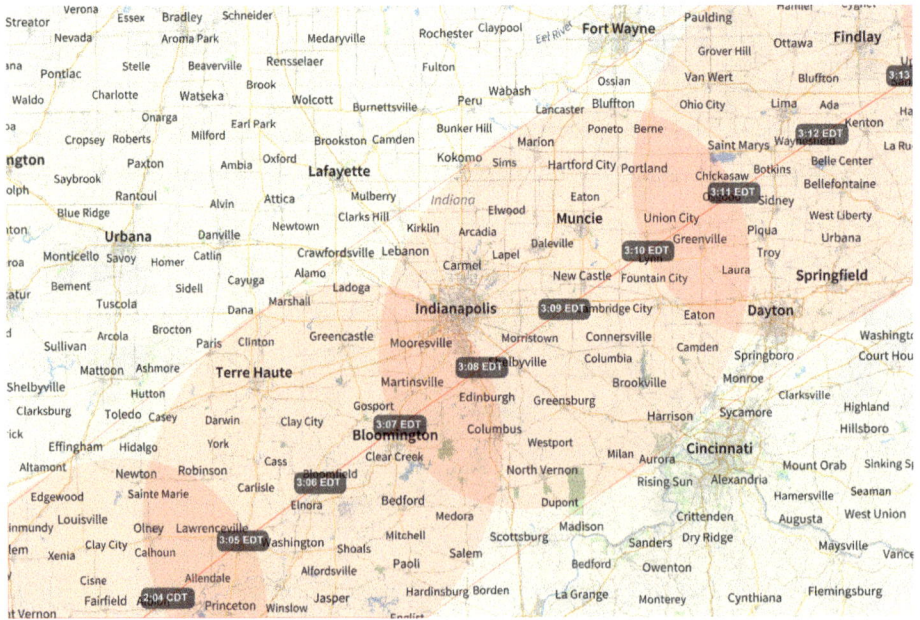

Detroit/Cleveland

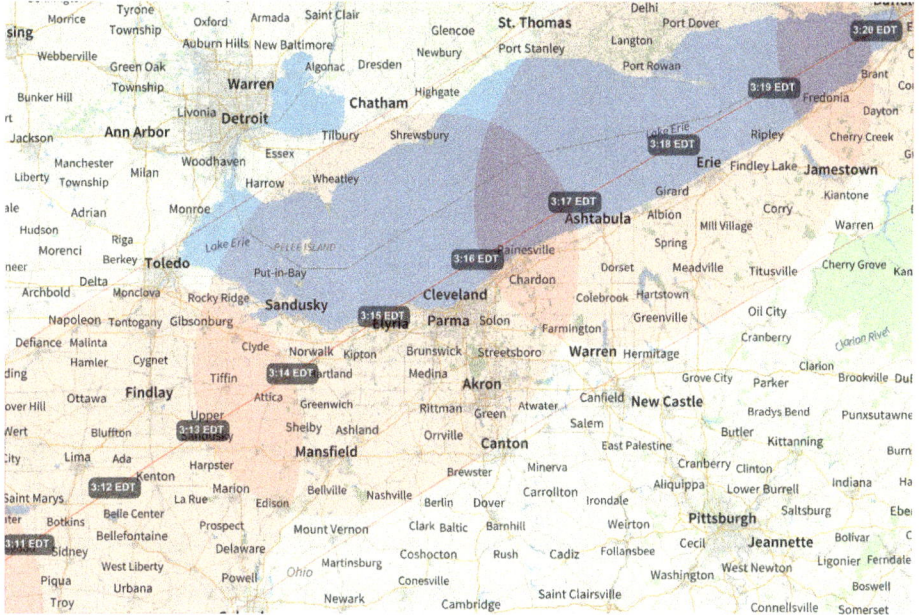

Buffalo/Niagara Falls

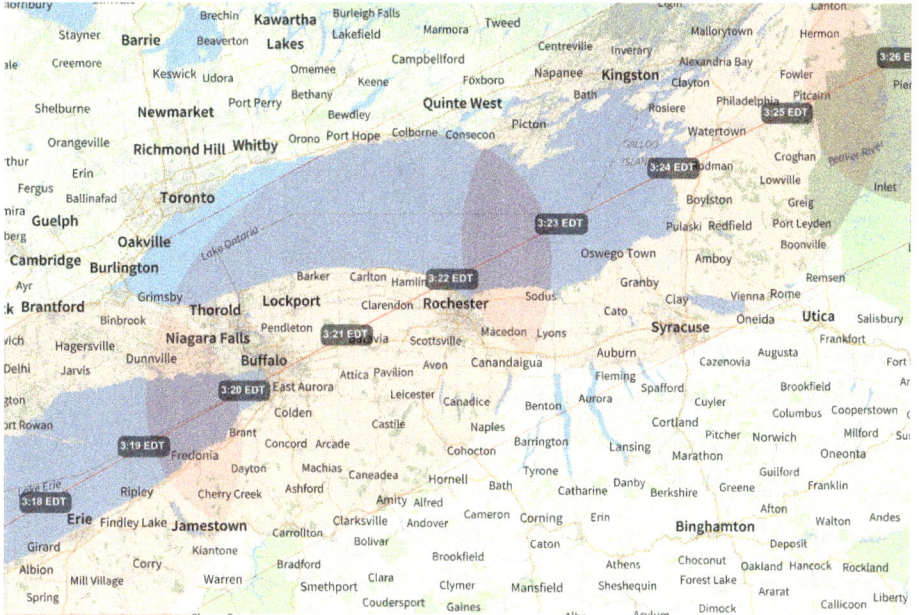

Table of Eclipses

Total Eclipse | April 8, 2024

Maximum duration of totality	4 minutes 32 seconds
Maximum totality nearest city	Nazas, Mexico
Maximum Sun coverage (by radius)	105.7%
Countries along totality path	Mexico, United States, Canada
Maximum duration of partiality	2 hours 41 minutes
Eclipse saros, inex designation	139, 55

Annular Eclipse | October 2, 2024

Maximum duration of totality	7 minutes 19 seconds
Maximum totality nearest city	Hanga Roa, Chile
Maximum Sun coverage (by radius)	93.3%
Countries along totality path	Chile, Argentina
Maximum duration of partiality	3 hours 29 minutes
Eclipse saros, inex designation	144, 63

Annular Eclipse | February 17, 2026

Maximum duration of totality	2 minutes 16 seconds
Maximum totality nearest city	(Antarctica)
Maximum Sun coverage (by radius)	96.4%
Countries along totality path	Antarctica
Maximum duration of partiality	2 hours 3 minutes
Eclipse saros, inex designation	121, 26

Total Eclipse | August 12, 2026

Maximum duration of totality	2 minutes 21 seconds
Maximum totality nearest city	Patreksfjörður, Iceland
Maximum Sun coverage (by radius)	104%
Countries along totality path	Greenland, Iceland, Spain
Maximum duration of partiality	2 hours 1 minute
Eclipse saros, inex designation	126, 34

Annular Eclipse | February 6, 2027

Maximum duration of totality	7 minutes 45 seconds
Maximum totality nearest city	Cidreira, Brazil
Maximum Sun coverage (by radius)	92.9%
Countries along totality path	Chile, Argentina, Uruguay, Brazil
Maximum duration of partiality	3 hours 27 minutes
Eclipse saros, inex designation	131, 42

Total Eclipse | August 2, 2027

Maximum duration of totality	6 minutes 26 seconds
Maximum totality nearest city	Luxor, Egypt
Maximum Sun coverage (by radius)	108%
Countries along totality path	Morocco, ..., Spain, Algeria, ..., Egypt, Saudi Arabia, Yemen, Somalia, ...
Maximum duration of partiality	2 hours 46 minutes
Eclipse saros, inex designation	136, 50

Annular Eclipse | January 26, 2028

Maximum duration of totality	10 minutes 20 seconds
Maximum totality nearest city	Camopi, French Guiana
Maximum Sun coverage (by radius)	92.2%
Countries along totality path	Peru, ..., Colombia, Brazil, ..., Spain, ...
Maximum duration of partiality	4 hours 3 minutes
Eclipse saros, inex designation	141, 58

Total Eclipse | July 21, 2028

Maximum duration of totality	5 minutes 14 seconds
Maximum totality nearest city	Kalumburu, Australia
Maximum Sun coverage (by radius)	105.7%
Countries along totality path	Cocos (Keeling) Islands, Christmas Island, Australia, New Zealand
Maximum duration of partiality	3 hours 5 minutes
Eclipse saros, inex designation	146, 66

Annular Eclipse | June 1, 2030

Maximum duration of totality	5 minutes 16 seconds
Maximum totality nearest city	Severnoye, Russia
Maximum Sun coverage (by radius)	94.5%
Countries along totality path	..., Turkey, ..., Russia, China, Japan
Maximum duration of partiality	3 hours 4 minutes
Eclipse saros, inex designation	128, 37

Total Eclipse | November 25, 2030

Maximum duration of totality	3 minutes 47 seconds
Maximum totality nearest city	Port-aux-Français, French Southern and Antarctic Lands
Maximum Sun coverage (by radius)	104.8%
Countries along totality path	Namibia, Botswana, South Africa, Lesotho, Australia
Maximum duration of partiality	2 hours 41 minutes
Eclipse saros, inex designation	133, 45

Annular Eclipse | May 21, 2031

Maximum duration of totality	5 minutes 18 seconds
Maximum totality nearest city	Minicoy, India
Maximum Sun coverage (by radius)	96%
Countries along totality path	..., Democratic Republic of the Congo, ..., India, ..., Thailand, ..., Indonesia
Maximum duration of partiality	4 hours 6 minutes
Eclipse saros, inex designation	138, 53

Hybrid Eclipse | November 14, 2031

Maximum duration of totality	1 minute 13 seconds
Maximum totality nearest city	Ua Huka, French Polynesia
Maximum Sun coverage (by radius)	101.1%
Countries along totality path	Panama
Maximum duration of partiality	3 hours 29 minutes
Eclipse saros, inex designation	143, 61

Annular Eclipse | May 9, 2032

Maximum duration of totality	17 seconds
Maximum totality nearest city	Gough Island, Saint Helena, Ascension and Tristan da Cunha
Maximum Sun coverage (by radius)	99.7%
Countries along totality path	(none)
Maximum duration of partiality	2 hours 48 minutes
Eclipse saros, inex designation	148, 69

Total Eclipse | March 30, 2033

Maximum duration of totality	2 minutes 39 seconds
Maximum totality nearest city	Utqiagvik, United States
Maximum Sun coverage (by radius)	104.7%
Countries along totality path	United States, Russia
Maximum duration of partiality	1 hour 55 minutes
Eclipse saros, inex designation	120, 24

Total Eclipse | March 20, 2034

Maximum duration of totality	4 minutes 13 seconds
Maximum totality nearest city	Biltine, Chad
Maximum Sun coverage (by radius)	104.7%
Countries along totality path	Nigeria, ..., Pakistan, China, India
Maximum duration of partiality	3 hours 2 minutes
Eclipse saros, inex designation	130, 40

Annular Eclipse | September 12, 2034

Maximum duration of totality	2 minutes 51 seconds
Maximum totality nearest city	Punta de Bombon, Peru
Maximum Sun coverage (by radius)	97.4%
Countries along totality path	Bolivia, Chile, Brazil, Argentina, ...
Maximum duration of partiality	3 hours 36 minutes
Eclipse saros, inex designation	135, 48

Annular Eclipse | March 9, 2035

Maximum duration of totality	42 seconds
Maximum totality nearest city	Mutuaura, French Polynesia
Maximum Sun coverage (by radius)	99.3%
Countries along totality path	New Zealand
Maximum duration of partiality	3 hours 11 minutes
Eclipse saros, inex designation	140, 56

Total Eclipse | September 1, 2035

Maximum duration of totality	2 minutes 58 seconds
Maximum totality nearest city	Northern Mariana Islands
Maximum Sun coverage (by radius)	103.3%
Countries along totality path	China, North Korea, Japan
Maximum duration of partiality	3 hours 1 minute
Eclipse saros, inex designation	145, 64

Total Eclipse | July 12, 2037

Maximum duration of totality	4 minutes 3 seconds
Maximum totality nearest city	Mount Isa, Australia
Maximum Sun coverage (by radius)	104.2%
Countries along totality path	Australia, New Zealand
Maximum duration of partiality	3 hours 12 minutes
Eclipse saros, inex designation	127, 35

Annular Eclipse | January 5, 2038

Maximum duration of totality	3 minutes 12 seconds
Maximum totality nearest city	Fernando de Noronha, Brazil
Maximum Sun coverage (by radius)	97.4%
Countries along totality path	..., Ghana, ..., Nigeria, Niger, Libya, Egypt, Chad, Sudan
Maximum duration of partiality	3 hours 50 minutes
Eclipse saros, inex designation	132, 43

Annular Eclipse | July 2, 2038

Maximum duration of totality	53 seconds
Maximum totality nearest city	Frontera, Spain
Maximum Sun coverage (by radius)	99.2%
Countries along totality path	Colombia, ..., Niger, Chad, Sudan, Ethiopia, South Sudan, Kenya
Maximum duration of partiality	3 hours 36 minutes
Eclipse saros, inex designation	137, 51

Total Eclipse | December 25, 2038

Maximum duration of totality	2 minutes 22 seconds
Maximum totality nearest city	Ōkārito, New Zealand
Maximum Sun coverage (by radius)	102.8%
Countries along totality path	Australia, New Zealand
Maximum duration of partiality	2 hours 51 minutes
Eclipse saros, inex designation	142, 59

Annular Eclipse | June 21, 2039

Maximum duration of totality	4 minutes 1 second
Maximum totality nearest city	Resolute, Canada
Maximum Sun coverage (by radius)	94.6%
Countries along totality path	United States, Canada, Russia, ..., Sweden, Estonia, Åland Islands, ...
Maximum duration of partiality	2 hours 25 minutes
Eclipse saros, inex designation	147, 67

Total Eclipse | December 15, 2039

Maximum duration of totality	1 minute 54 seconds
Maximum totality nearest city	(Antarctica)
Maximum Sun coverage (by radius)	103.6%
Countries along totality path	Antarctica
Maximum duration of partiality	1 hour 46 minutes
Eclipse saros, inex designation	152, 75

Total Eclipse | April 30, 2041

Maximum duration of totality	1 minute 55 seconds
Maximum totality nearest city	Luanda, Angola
Maximum Sun coverage (by radius)	102%
Countries along totality path	Angola, Democratic Republic of the Congo, Uganda, Kenya, ...
Maximum duration of partiality	3 hours 13 minutes
Eclipse saros, inex designation	129, 38

Annular Eclipse | October 24, 2041

Maximum duration of totality	6 minutes 1 second
Maximum totality nearest city	Enewetak, Marshall Islands
Maximum Sun coverage (by radius)	94.7%
Countries along totality path	China, Mongolia, North Korea, Japan, ...
Maximum duration of partiality	3 hours 37 minutes
Eclipse saros, inex designation	134, 46

Total Eclipse | April 19, 2042

Maximum duration of totality	4 minutes 55 seconds
Maximum totality nearest city	Nishinoomote, Japan
Maximum Sun coverage (by radius)	106.2%
Countries along totality path	Indonesia, Malaysia, Brunei, Philippines
Maximum duration of partiality	2 hours 41 minutes
Eclipse saros, inex designation	139, 54

Annular Eclipse | October 13, 2042

Maximum duration of totality	7 minutes 38 seconds
Maximum totality nearest city	Mount Isa, Australia
Maximum Sun coverage (by radius)	93.1%
Countries along totality path	Thailand, Malaysia, Indonesia, East Timor, Australia, New Zealand
Maximum duration of partiality	3 hours 30 minutes
Eclipse saros, inex designation	144, 62

Total Eclipse | April 9, 2043

Maximum duration of totality	1 minute 49 seconds
Maximum totality nearest city	Talaya, Russia
Maximum Sun coverage (by radius)	104.2%
Countries along totality path	Russia
Maximum duration of partiality	1 hour 44 minutes
Eclipse saros, inex designation	149, 70

Annular Eclipse | October 2, 2043

Maximum duration of totality	2 minutes 12 seconds
Maximum totality nearest city	(Antarctica)
Maximum Sun coverage (by radius)	94.4%
Countries along totality path	(none)
Maximum duration of partiality	2 hours 5 minutes
Eclipse saros, inex designation	154, 78

Annular Eclipse | February 28, 2044

Maximum duration of totality	2 minutes 24 seconds
Maximum totality nearest city	(Antarctica)
Maximum Sun coverage (by radius)	96.1%
Countries along totality path	South Georgia and the South Sandwich Islands
Maximum duration of partiality	2 hours 0 minutes
Eclipse saros, inex designation	121, 25

Total Eclipse | August 22, 2044

Maximum duration of totality	2 minutes 6 seconds
Maximum totality nearest city	Fort Simpson, Canada
Maximum Sun coverage (by radius)	103.7%
Countries along totality path	Greenland, Canada, United States
Maximum duration of partiality	1 hour 55 minutes
Eclipse saros, inex designation	126, 33

Annular Eclipse | February 16, 2045

Maximum duration of totality	7 minutes 41 seconds
Maximum totality nearest city	Avarua, Cook Islands
Maximum Sun coverage (by radius)	92.9%
Countries along totality path	New Zealand, Cook Islands
Maximum duration of partiality	3 hours 26 minutes
Eclipse saros, inex designation	131, 41

Total Eclipse | August 12, 2045

Maximum duration of totality	6 minutes 9 seconds
Maximum totality nearest city	Freeport, Bahamas
Maximum Sun coverage (by radius)	107.8%
Countries along totality path	United States, ..., Haiti, ..., Venezuela, Trinidad and Tobago, Guyana, ..., Brazil
Maximum duration of partiality	2 hours 42 minutes
Eclipse saros, inex designation	136, 49

Annular Eclipse | February 5, 2046

Maximum duration of totality	9 minutes 35 seconds
Maximum totality nearest city	Pyramid Point, Kiribati
Maximum Sun coverage (by radius)	92.4%
Countries along totality path	Indonesia, Papua New Guinea, Solomon Islands, ..., United States
Maximum duration of partiality	3 hours 55 minutes
Eclipse saros, inex designation	141, 57

Total Eclipse | August 2, 2046

Maximum duration of totality	4 minutes 56 seconds
Maximum totality nearest city	Longonjo, Angola
Maximum Sun coverage (by radius)	105.4%
Countries along totality path	Angola, Namibia, Botswana, South Africa, Eswatini, French Southern and Antarctic Lands
Maximum duration of partiality	3 hours 4 minutes
Eclipse saros, inex designation	146, 65

Annular Eclipse | June 11, 2048

Maximum duration of totality	4 minutes 53 seconds
Maximum totality nearest city	Djúpivogur, Iceland
Maximum Sun coverage (by radius)	94.5%
Countries along totality path	United States, Canada, ..., Russia, ..., Uzbekistan, Turkmenistan, Afghanistan
Maximum duration of partiality	2 hours 51 minutes
Eclipse saros, inex designation	128, 36

Wolfram Precision Eclipse Website

Get a precision report on what to expect from your exact viewing location.

www.precisioneclipse.com

Index

www.ingramcontent.com/pod-product-compliance
Lightning Source LLC
Chambersburg PA
CBHW060620200326
41521CB00007B/828